THE
WAKE-UP
CALL

THE

WHY THE PANDEMIC HAS EXPOSED THE

WAKE-UP

WEAKNESS OF THE WEST, AND HOW TO FIX IT

CALL

JOHN MICKLETHWAIT and **ADRIAN WOOLDRIDGE**

HarperVia

An Imprint of HarperCollinsPublishers

HarperCollins books may be purchased for educational, business, or sales promotional use. For information, please email the Special Markets Department at SPsales@harpercollins.com.

FIRST EDITION

Designed by Yvonne Chan

Library of Congress Cataloging-in-Publication Data is available upon request.

ISBN 978-0-06-306529-1

20 21 22 23 24   LSC   10 9 8 7 6 5 4 3 2 1

CONTENTS

INTRODUCTION:
WEEKS WHEN DECADES HAPPEN

n 1651, a gentleman-scholar who readily admitted that "fear and I were born twins," published one of the great books on government. Thomas Hobbes had survived the notoriously bloody English Civil War by fleeing to France—and his great philosophical concern was personal safety. Life in a state of nature was "solitary, poor, nasty, brutish and short" because people were always fighting each other. So, he argued, citizens should form a contract to give up their freedoms to a ruler who could offer them protection. The idea that the state's legitimacy depended on keeping its citizens safe was revolutionary. Back then, kings claimed their position came by divine right, not contract. Hobbes, who also managed to survive the Great Plague in 1665–66 and died in his bed aged ninety-one, chose for the book's frontispiece a picture of a single great ruler composed of the bodies of hundreds of tiny little subjects. He called it *Leviathan*.

If Hobbes were resurrected today, he would feel vindicated. The Covid-19 pandemic has unleashed fear into the world on a scale not seen since the Second World War. By the end of June

2020, ten million people had been infected worldwide and more than five hundred thousand had died, a quarter of them in the United States. Every day brought horrific images: New York City paying prisoners in hazmat suits to help dig makeshift graves for piles of wooden coffins; Britain's prime minister fighting for his life in intensive care; Médecins Sans Frontières setting up camp in the center of Brussels; the president of the United States suggesting people inject themselves with disinfectant. Fear of death has been accompanied by fear of economic ruin. With the total cost to the world economy running into trillions of dollars, whole industries have closed down. Millions of people who thought that their livelihoods were secure now rely on government checks.

Indeed, everywhere you look, people who once did not care much about the state have been turning to Leviathan to protect them—and, as Hobbes predicted, giving up their most cherished liberties, even the freedom to leave their own homes. We have allowed the state to regulate our every move. And when the state has let us down, we have broken into fury—most obviously on the streets of America, after the killing of George Floyd.

The Coronavirus has made government important again. Not just powerful again (look at those once-mighty companies begging for help), but also vital again. It matters enormously whether your country has a good health service and competent bureaucrats. The arrival of the virus was like an examination of state capacity. A handful of Western countries passed. Germany was an outstanding performer in Europe, while Denmark, Norway,

Switzerland, and, surprisingly, Greece did well. New Zealand and Australia were champions on the Pacific rim. But most Western countries, particularly America and Britain, failed the test, humiliatingly so when compared with countries in Asia.

The numbers underline this.[1] By midyear, the death rate in Belgium was 850 per every million people; in Britain it was above 650; Italy and Spain were both around 550, while the United States, where the virus was once again surging in Arizona, Texas, and Florida, was closing in on 400. The figure for Germany was around 100. In South Korea and Japan the death rate was just seven and five respectively. Mainland China claims a figure of three. That final number comes with a lot of caveats, but even if the Chinese death toll was in fact ten times the official total, the regime would still be ten times better at protecting its people than Donald Trump was. Even relatively poor parts of Asia, such as Vietnam and Kerala state in India, outperformed both the United States and Britain by dramatic margins.

Looking at individual cities, the comparison between the West and Asia is even starker. London and New York City are both a little smaller than Seoul. But by the end of June 2020, when New York City had seen twenty-one thousand Covid-related deaths and London six thousand, the Korean capital had lost just six people. Seoul's politicians united quickly and put testing booths in many streets, while in New York the politicians feuded with each other and Trump, and people joked that the only way for a poor New Yorker to get a Covid test was to cough near a rich one. In Seoul's hospitals, doctors had all the necessary equipment; in New York doctors wore ski masks that they bought for themselves, nurses dressed in garbage bags, and infected and noninfected patients were mixed up together, helping to spread the disease.[2] In London, more people died of the virus in one four-week stretch in April than in the worst four weeks of the Blitz.[3] By early June, Seoul had ended its lockdown and was going back to work, while New York was thronged with mask-wearing protesters, brought on the streets by Floyd's killing.

"The crisis demanded a response that was swift, rational, and collective," George Packer lamented in *The Atlantic*. "The United States reacted instead like Pakistan or Belarus—like a country with shoddy infrastructure and a dysfunctional government whose leaders were too corrupt or stupid to head off mass suffering."[4] America's failures at home coincided with a failure to mobilize the Western alliance. Since the Second World War, the United States has led the West. Not so during this pandemic. Trump sidelined global organizations and squabbled with his allies, failing to inform them that he was banning flights from Europe. Not that the European Union covered itself with glory: it failed to help Italy and Spain in their darkest hours, and then started bickering about a stimulus plan.

A REVERSAL OF FORTUNE

Squabbling in the West gave China an opportunity. Medical equipment was dispatched to Italy in boxes printed with the lyrics of Italian operas and to Hungary printed with its leader Viktor Orban's favorite slogan, "Go Hungary." "The situation I see can be described as this," Orban declared in an interview relayed on Chinese state television, "In the West, there is a shortage of basically everything. The help we are able to get is from the East."[5] When the Floyd protests erupted, with four in five Americans thinking their country was "spiraling out of control," the Chinese sarcastically compared protesters in Minneapolis and other cities to the pro-democracy ones in Hong Kong.[6] On the last day of June, with American officials confessing that the virus was "going in the wrong direction," Boris Johnson having to re-lockdown the city of Leicester, and Emmanuel Macron preparing to sack his entire government in Paris, China felt confident enough to impose a harsh new security law on Hong Kong. By then, a virus that in January had looked as if it might be "China's Chernobyl" looked more like the West's Waterloo.

When Hobbes wrote *Leviathan*, China was the center of administrative excellence. It was the world's most powerful country with the world's biggest city (Beijing had more than a million inhabitants), the world's mightiest navy, and the world's most sophisticated civil service, run by scholar-mandarins chosen from across a vast empire by rigorous examinations. Europe was a bloodstained battlefield ruled by rival feudal families, where government jobs were either allotted by birth or bought and sold like furniture. Gradually, Europe's new nation-states overtook the Middle Kingdom because they underwent three revolutions unleashed by national rivalries and political ideas.

The first revolution, which Hobbes helped inspire, was the creation of the competitive nation-state. Europe's monarchies simultaneously imposed order on barons at home while vying with each other for supremacy abroad. When the Chinese invented gunpowder they used it for fireworks; Europeans used it to blow one another (and then the Chinese) out of the water. The second leap forward was the leaner liberal state that emerged after the American and French Revolutions. Thinkers such as Adam Smith and John Stuart Mill argued that corrupt monarchical privilege should give way to meritocratic efficiency. In Victorian Britain, liberals built the schools, hospitals, and sewers that we still use and paid for them by cutting wasteful spending on sinecures and aristocratic fops (staggeringly, the nation's tax bill fell from just under £80 million in 1816 to well under £60 million in 1846).[7] The third big change was the arrival of the welfare state just over a century ago. The theme of the welfare state was security once again—but this time it meant health, education, and pensions, not just protecting you from your brutish neighbors, and the definition has kept expanding ever since. William Beveridge, the architect of Britain's New Jerusalem in 1945, wanted his Leviathan to slay the "five giants" of Want, Disease, Ignorance, Squalor, and Idleness.

While the West went through this frenzy of invention, the East slept. Both the Ottomans and Mughal India, which had looked so superior in Hobbes's day, crumbled from within. China turned inwards: its government concentrated on regulating the minutiae of daily life (one of its six departments focused on ceremonies and etiquette), while would-be civil servants were tested on the Confucian classics, not modern economics. The empire collapsed in the early twentieth century, ushering in an era of instability that culminated, at its most horrific, in Mao's Cultural Revolution. By the 1960s, a complete reversal had taken place, with the gap between the West and the East as wide as it had been in Hobbes's day. America was putting a man on the moon while millions of Chinese were dying of starvation.

Since then two things have happened. First, the Western state has atrophied. The mid-1960s was not only the last time the public sector was on a par with the private sector; it was the last time that people in many countries trusted their government. Leviathan overreached, promising more than it could deliver; the 1970s brought stagflation, an oil crisis, and Watergate. In the 1980s, Ronald Reagan and Margaret Thatcher launched a counter-revolution that spread around the world, even reaching socialist bastions like Sweden. But they were much more successful in changing the rhetoric than the reality, so the state has continued to grow; only now it is a much more loathed monster.

Then along came the populists. Silvio Berlusconi was the trailblazer, promising to boost his fellow Italians' fortunes but mysteriously only boosting his own (Italy's economy under his rule grew more slowly than any other country, except Zimbabwe and Haiti). In 2016, Donald Trump appeared, vowing to "drain the swamp" in Washington, DC, while Britain voted for Brexit. Four years later, the swamp is fuller than ever, and Britain is in danger of leaving the European Union chaotically. On one day in June more people died of Covid in Britain than in the whole of the EU.[8] The two countries that have set the mood music for the West for the past half century, look divided and shambolic.

Second, Asia has rebuilt. For a while all the West noticed was Japan, but its success was economic rather than political (and the economic bubble burst). At the same time, Singapore was creating a new model of government. A colonial satrap can now claim to be the best-governed small state in the world, with better schools and hospitals than any Western country. Like the Victorians, Singapore has achieved this not by spending large sums of money—it spends less than 20 percent of GDP on government—but by taking government seriously. It pays its top civil servants spectacularly well, but weeds out poor performers, notably bad teachers. South Korea and Taiwan are following the same path. China is more of a mixed picture. Xi Jinping's decision to declare himself leader-for-as-long-as-he-wants in 2018 smacks more of feudalism than late modernity. But China is also following the Singaporean model of trying to create a world-beating government, led by a caste of elite mandarins. Its schools are racing up the league tables.

Thus the East's success with Coronavirus is not a lucky accident: it is the result of a change that has been several decades in the making. Asia had the technology to deal with the disease—especially "intelligent cities" that use smart infrastructure to manage urban life better. It had the trust of its citizens. Is it any surprise that Asian schools do so much better than America's when the school calendar in cities like New York is still organized on the assumption that children need a long summer break to collect the harvest? Anybody who flies from shoddy Kennedy to one of Asia's gleaming new airports can see an infrastructure gap yawning. The danger is that the same gap is opening with the state more generally.

The pandemic is not the only exam the West is failing. Climate change could do even greater damage than the virus in the long run, but Trump is blocking serious action while European leaders follow Saint Augustine's plea of "make me chaste but not yet." Angela Merkel has been pointing out for years that Europe accounts for 7 percent of the world's population and a quarter of its economy but half its social spending—without her, or the EU, doing anything about it. In America, Floyd's killing was part of a pattern of inequality that stretches back decades. Many Black Americans live in a Hobbesian world, where Leviathan is a source of fear rather than security.

WEEKS WHEN DECADES HAPPENED

Lenin once said that "there are decades where nothing happens; and there are weeks when decades happen." This virus, like many before it, is just such a history-accelerating crisis. Democratic Athens gave way to militaristic Sparta after a horrific plague killed a third of its population, convulsing its institutions and enfeebling its army. (Hobbes was the first person to translate Thucydides's great history of this tragedy.) Two great plagues, Cyprian and Justinian, helped bring down the Roman Empire.[9]

The virus is a wake-up call. If we respond to the Covid crisis intelligently, treating it as not just a public health disaster but a stress test of Western government, if we take government as seriously as Asia does, we can preserve the Western advantage that began in Hobbes's time. If we ignore the wake-up call, or hit the snooze button, a Sino-centric world beckons, with the United States becoming a large offshore island, while Europe returns to what it was five centuries ago—an archipelago at the poorer, western end of Eurasia.

The West is more than a geographical expression. It is an idea that has freedom and human rights at its heart. The West has often fallen behind its own standards, not least on race, but at least it has those standards. The Chinese Communist Party cares nothing for democracy or civil rights, as the citizens of Hong Kong and the Uighurs can testify. The aim of this short book is to explain to people why government matters, especially in promoting liberty and democracy, and to lay out a manifesto for reform that would rebuild the West and expand it, so that it becomes the center of a union of global democracies.

Our first two chapters deal with the history of the state up until Covid, because you can only understand the present crisis in the light of the past (something China now studies harder than we do). We tell the story, first of how the West pulled ahead of the rest of the world by embracing new ideas and technology, and then how, since the 1960s, the West has neglected statecraft, while the East has regained its focus. In Chapter Three, we look at what is wrong with Western government—why it is both overloaded and unloved, a grumpy compromise between taxpayers who think it costs too much and consumers who think it delivers too little. In Chapter Four, we look at the great Covid test—and why too many parts of the West failed it. Then we ask: how can we change this?

Our fear, outlined in Chapter Five, is that the West will choose the wrong solutions: blaming it all on the admittedly incompetent populists, or deciding that the answer to an overloaded state is to add even more government. The worst result would be if people around the world embraced strongmen and autocracy. Even if China has generally done better with Covid than America, many democracies have done better than China—while Vladimir Putin's popularity is at its lowest since he came to power and Belarus's thuggish ruler may have out-Trumped Trump by declaring that the best way to fight the virus is to drink vodka and drive a tractor.

We will argue that the West needs better government, not more of it. This is partly a question of pragmatic modernization. In Chapter Six, we conduct a thought experiment by combining the two greatest liberal leaders of the nineteenth century—William Gladstone and Abraham Lincoln—and imagine what President Bill Lincoln could do if he was installed in the White House, merely by using what works in other countries. Our President Lincoln would, among other things, raise the retirement age, provide nearly free health care for all, pay some civil servants million-dollar salaries, introduce national service (civic rather than military), decimate the tax code, sack bad teachers, stop selling ambassadorships, wage war on the old corruption that diverts so much cash to the wealthy—and trumpet freedom around the world. Bill Lincoln's West would be broader and more inclusive. The result at home would be a bigger state in some areas, but a far smaller one in others.

But you can't avoid political thought. Even if single-payer systems like Canada and Sweden deliver better health care at a lower cost to the taxpayer (let alone the consumer), many Americans will still claim they don't want "socialism." Tell Britons that the NHS would have saved more lives if it had more use of the private sector, and they will assume you want to sell it off. Suggest to the French that pensions should rise with life expectancy if they want to avoid bankruptcy, and they mount the barricades. So at some point, you have to start asking the same question that Hobbes did: what is the state for?

In our conclusion, we return to the issues of security and liberty that Covid has again reinforced. For the moment security is ahead: during a pandemic the health of the most privileged is dependent on the health of the least privileged, so the fight against a virus is necessarily a collective one. But you cannot build a state just on that basis. The security state can suppress both freedom and creativity (Hobbes's old university, Oxford, burned all his books, including *Leviathan*, in the quadrangle of the Bodleian Library in 1683). The West's competitive advantage over China—the modern book-burner (or website-blocker)—lies in that freedom, and the entrepreneurialism it unleashes. If the current test is about security, the next one will be about liberty. That is a battle that the democracies of the world need to win together, and they will do so much more easily if they line up behind a common vision of freedom.

THE ALARM CLOCK

The secret of the West's success over the past four hundred years is its appetite for creative destruction: just when everything looks hopeless, it succeeds in regenerating itself. Look at the changes wrought by the Roosevelts in the Gilded Age and the Great Depression. Having almost destroyed itself during the Second World War, Europe became the continent of *les trentes glorieuses* and *wirtschaftswunder*. Now the West has to recover from Covid, at a time when America's power is waning and China is strengthening. The question is whether the West can rise to the challenge as it has so many times before and rethink the theory and practice of government—or whether it will fumble about, letting liberty slip away and leaving China to reclaim the global leadership it had when that frightened old tutor sat down to write *Leviathan*.

It is time to wake up.

CHAPTER ONE

THE
RISE OF
THE WEST

n the early fourteenth century, Siena was about as close as
you could get to the heart of Western civilization. Prosperous,
progressive, and, thanks to the budding Renaissance, a center
of art and intellect, the Italian city was ruled by a council of nine
elected officials, who commissioned Ambrogio Lorenzetti to paint
an Allegory of Good and Bad Government for the walls of their
chamber. He finished the fresco in 1339. On the Good Government
side, the shops are open, builders are at work, people are danc-
ing, and Justice is a beautiful woman, guided by God. "Turn your
eyes to behold her," the inscription below implores the council
members. "Look how many goods derive from her and how sweet
and peaceful is that life of the city where is preserved this virtue

who outshines any other." An inspiring notion till you look at the Bad Government side. The fresco there is much more damaged—appropriately enough—but the image is still clear: Justice lies bound at the feet of the Tyrant, with the characters of Cruelty, Deceit, Fraud, Fury, Division, and War looking on.

Nine years after the fresco was finished, Siena was ravaged by the Black Death (which killed a third of the population of Europe) and the Council of Nine was ousted. The city became a battleground of competing factions, frequently backed by outsiders, including the Papacy and France. It was not until the early sixteenth century that Siena achieved some degree of stability again, and then only by bowing down to a tyrant, Pandolfo Petrucci. One of Petrucci's visitors was a cynical Florentine ambassador called Niccolò Machiavelli.

Italy's small city-states were unusually fratricidal and unstable, but even two hundred years after Lorenzetti put down his brush, politics across Europe resembled the Bad Government rather than the Good Government side of his fresco. Everywhere rival aristocratic houses were feuding over the chance to become Tyrants themselves, enlisting Cruelty, Deceit, and Fraud on their side. Most countries had been through civil wars, such as the Wars of the Roses in England. And after Martin Luther published his ninety-five theses in 1517, the Protestant Reformation added an even more profound reason for societies to tear themselves apart.

There were, of course, a few people who dreamed about good governance. Desiderius Erasmus, Luther's contemporary, dreamed of a *via media* that would unite Europe behind a reformed church and enlightened rulers. He hung around the great courts of the day preaching the virtues of civic reform. His *Education of a Christian Prince* (1516), a how-to-rule book addressed to Charles of Spain, the future Habsburg emperor, argued that princes should have one great object in life, to serve the public good. "Only those who dedicate themselves to the state, and not the state to themselves deserve the title 'prince'." He was particularly scalding about the "long-standing and terrible mania among Christians for war."[1] A noble sentiment that evaded Charles V as he put much of Europe to the sword.

In her wonderful series of novels about Thomas Cromwell, Hilary Mantel imagines Henry VIII's first minister daydreaming about Lorenzetti's Allegory while his monarch muses about the perfidy of wives.[2] It is October 1536, and Cromwell, the blacksmith's son hailed as the great Protestant modernizer of Tudor government, is cock of the walk. But even then, leading nobles are suspicious of the upstart, Cromwell is terrified of annoying the king, and most of the north of England is in revolt on "a Pilgrimage of Grace." Four years later, Cromwell was executed, in large part because the new wife he had found for Henry, Anne of Cleves, was not as pretty as he had advertised.

Looked at in global terms, Christendom was a sideshow. While Cromwell was struggling to run a country of perhaps 2.5 million people from London, Suleiman the Magnificent ruled over ten times that number from Istanbul. The Ottoman Empire stretched from modern Hungary to the Middle East and North Africa, with a single legal code and free education for Muslim boys. In India, the Mughal Empire was under construction. And beyond that lay China, a country the size of Europe, ruled over by a single emperor, held together by a network of roads and canals and the world's best bureaucracy. The imperial quarter in Beijing, where the royal family and its functionaries lived, housed a population roughly the same size as London. Anybody who entered the presence of the emperor demonstrated their awe of the "son of heaven" by kowtowing—kneeling and touching their head nine times on the ground. If Henry and Cromwell had come to Beijing, they would have been treated as primitive curiosities.

The great explorers that Europe sent forth were often only rediscovering what Muslim and Asian traders had known for centuries. When Vasco da Gama boldly rounded the Cape of Good Hope in 1497, "opening up" a route to India, the Muslim rulers he met in East Africa scorned his paltry gifts but generously—or condescendingly—lent him navigators who had been plying the route to India for years. Da Gama's entire fleet of four tiny ships could just about have been squeezed into one of China's vast treasure ships that the Chinese admiral Zheng He had brought to East Africa earlier that century.[3]

Yet history was turning. The Ottomans were held at Vienna in 1529—and never got any further west. While da Gama was one of the first of a flotilla of European explorers, Zheng He was the last of his breed. In 1433 the emperor banned all further voyages overseas and ordered the destruction of all oceangoing ships. Successive emperors claimed that the rest of the world didn't have anything to teach the Celestial Kingdom—and focused their formidable powers on regulating the minutiae of domestic life, including the size of your house. "Though [the emperor] may live in the deepest retreat of his palace, at the end of tortuous corridors," one courtier wrote, "nothing escapes him, nothing is hidden from him, nothing can escape his vigilant watch."[4] China became a "homeostatic society," held back also by Confucianism, which claimed that an ancient thinker had solved all the world's problems (the same applied to the Koran in the Islamic world).[5] In 1792, when Lord George Macartney arrived in Beijing bearing telescopes, clocks, a barometer, a spring-suspension coach, and other wonders, the emperor, Qianlong, famously first kept the British envoy waiting for months and then dismissed the gifts with a contemptuous wave. "We have never valued ingenious articles, nor do we have the slightest need of your country's manufactures."[6]

In fact, with China absent and the Ottomans asleep, European armies and ideas were conquering the planet. By the nineteenth century, London, which seemed so insignificant in Henry VIII's time, had become the world's most important city. In Anthony Trollope's *The Way We Live Now* (1875), the emperor of China appears as a curiosity at a grand dinner given by a crooked financier. "He sat there for more than two hours, awful, solid, solemn, and silent, not eating very much—for this was not his manner of eating; nor drinking very much—for this was not his manner of drinking; but wondering, no doubt, within his own awful bosom, at the changes which were coming when an emperor of China was forced, by outward circumstances, to sit and hear this buzz of voices and this clatter of knives and forks."[7]

China drifted even further away from the center of power in the twentieth century. In the West, the United States took over from Britain as the heart of the global economy, Germany made two bloody bids for continental supremacy, and European empires were dismantled. Still, no matter what the West did to damage itself, China remained behind.

HOBBES AND THE NATION-STATE

Why did Europe forge ahead? Historians have suggested countless answers, citing everything from geography to pathogens. But better government was a big reason. Three great revolutions gave rise successively to the nation-state, the liberal state, and the welfare state. In each case new ideas reflected and reinforced changes that were taking place in society. We will examine these ideas through three great thinkers: Thomas Hobbes, John Stuart Mill, and Beatrice Webb. But it is striking how the spur that pushed first Europe and then the West forward was competition.

Cromwell may have lost the battle to the barons—and with it his head—but from the point of view of the nation-state, he won the war. With bloody interruptions, the crown gradually extended its authority over the rest of the country. The same happened across Europe. In Protestant countries, the Reformation gave princes direct power over the church, while Catholics, like Charles V, used the Counter-Reformation to increase their clout.[8] In Spain and France, Count Olivares and Cardinal Richelieu built great administrative machines. But even Europe's mightiest monarchs never had the absolutist power of China's emperor; they had to cooperate with parliaments, corporations, and the law courts, and they were always in competition with each other.

This could be bloody. There were only three years in the seventeenth century that were free from wars between states (1610, 1670, and 1682). In the Thirty Years' War (1618–48), Germany alone lost around a third of its population before the Treaty of Westphalia allowed local rulers to dictate their peoples' religion ("*cuius regio eius religio*").[9] But conflict also drove innovation: "War made the state and the state made war," as one historian puts it.[10] Desperate to discover soldiers who understood artillery and supply chains, monarchs recruited new men to supplement the old officer class.[11] Desperate to fill their treasuries, they handed out charters to merchants to forge trading routes. Thus, tiny Portugal ended up ruling Brazil, while Holland gained control of Indonesia.

Britain was very much at the center of all this tumultuous state-building. Oliver Cromwell (a distant relation of Thomas) led a revolution that established a commonwealth in 1649–60, with parliament declaring kingship to be "unnecessary, burdensome, and dangerous to the liberty, safety and public interest of the people."[12] Britain's oligarchs welcomed science and ideas. Its soldiers and sailors worked closely with its merchants and their new companies to expand abroad.

This was the world in which *Leviathan* was born. Thomas Hobbes was only one of several European thinkers, including Machiavelli in Italy and Jean Bodin in France, to ask probing questions about the state. But *Leviathan* represented a different level of intellectual sophistication. Hobbes was the first person to argue that the *right* to rule depended neither on the will of God (the Christian view) nor on the clever use of force (Machiavelli's heresy) but on a contract between the ruler and the ruled.

Hobbes's starting point was the bloodstained Europe where Justice lay at the Tyrant's feet and every chief minister feared the chopping block. Even his birth was hastened by fear—he was born prematurely in 1588 because his mother worried that the Spanish Armada had landed. He scraped together a living as a tutor—and eventually fled to Paris in 1640 to escape the coming civil war. The publication of *Leviathan* in 1651 (two years after the execution of Charles I) led to another rushed departure—this time back to England to escape royalists who hated *Leviathan*'s anticlerical tone. That, in turn, put Hobbes at the mercy of Oliver Cromwell. Only when his former pupil was restored to the throne as Charles II in 1660 and gave him a pension did he feel remotely safe.

This history explains the starting point for *Leviathan*: that we need a powerful state to prevent us from trying to kill each other, which is what happens in "a state of nature." For Hobbes, we are all little atoms of appetite and fear. Left to our own devices, we will try to outfox or outfight each other and reduce the world to a "war of every man against every man." To avoid that, we should surrender our freedoms to Leviathan, who keeps law and order.[13] The only right we get to keep is the right to save our own lives if they are directly threatened.

Hobbes infuriated everybody—radicals because he gave so much power to the state, royalists because he thought that the state's legitimacy only came from the people (and he did not mind whether the security-provider came in the shape of a king or a parliament). Yet *Leviathan* fitted a world where nation-states were becoming more powerful—the number of sovereign bodies in Europe shrank from around four hundred at the end of the Middle Ages to about twenty-five at the beginning of the First World War[14]—but also a world where individuals were gaining freedom to make their fortunes and question old orthodoxies. By the time Hobbes died (in his bed) in 1679, political thinking was coming alive.[15] Liberal thinkers such as John Locke, who went up to Oxford the year after *Leviathan* was published, and later Adam Smith and David Hume, all sought to limit the state's power. Thomas Paine, the author of *Common Sense* (1776) and *The Rights of Man* (1791), was even ruder: "Society is produced by our wants and government by our wickedness; the former promotes our happiness positively by uniting our affections, the latter negatively by restraining our vices."[16] On the continent, Jean Jacques Rousseau's *Social Contract* (1762) wanted the state to be controlled by the "general will."

These ideas helped create two great upheavals, in America in 1776 and in France in 1789. Having thrown out the British, the Founding Fathers drew heavily on Locke's version of the social contract—though Thomas Jefferson changed his "life, liberty, and property" into the more poetic "life, liberty, and the pursuit of happiness." Parts of Rousseau's version of the social contract were copied word for word into the French revolutionaries' Declaration of the Rights of Man and the Citizen in 1789. Meanwhile, industrialization had just started in Britain. Suddenly there were radical thoughts everywhere—that all men were created equal; that reason and science, not religion and tradition, should guide government; that individual happiness was the *summum bonum*. But France collapsed into the Terror, Napoleon, and the Bourbons, and young America, although influential, was small and far away. So the next revolution in government—and perhaps the one that is most relevant to the West's current dilemma—took place in what had become the world's most powerful country.

MILL AND THE LIBERAL STATE

There were two ways to look at Britain after the battle of Waterloo in 1815. One was as the world's only superpower: the vanquisher of Napoleon, the mistress of the seas, the center of the global economy, and the ruler of an empire that stretched from India to North America. The other was as a sprawling, inefficient, antiquated sewer of privilege and corruption—an *ancien régime* that needed to be radically modernized using all the tools of reason and reform so that its government matched its commercial power.

In 1815, William Gladstone and John Stuart Mill, who would become the two great lions of Victorian liberalism, were growing up on opposite sides of this divide. Mill was only nine years old when Waterloo was won, and Gladstone only six. But they were both prodigies. Gladstone supposedly delivered his first political speech when he was three (he was brought into a dinner his father was throwing for the future Conservative prime minister, George Canning, and proclaimed "ladies and gentlemen"), while Mill was even more of a prig, having learned Greek by the time he was three and Latin by eight.

Gladstone was brought up a Tory—a protector of the status quo. As a young man, he denounced the Great Reform Bill and, to his later shame, defended his father's record as an owner of slaves in the West Indies. The historian Thomas Macaulay described him as "the rising hope" of the "stern and unbending Tories."[17] Mill's background was more enlightened. His father, James Mill, was a believer in "liberty," "reason," and "effort," all of which were being frustrated by the establishment—and he raised his son to be "a reformer of the world."[18] John Stuart's godfather was Jeremy Bentham, who pioneered the utilitarian idea that every institution should be measured by how well it advanced the greatest happiness of the greatest number, and he was surrounded by radicals such as David Ricardo, the economist who invented the notion of "comparative advantage," and John Wade, the compiler of The Extraordinary *Black Book*, which listed all the nepotistical abuses of government, rotten boroughs, sinecures, and all.

John Stuart Mill's focus was on ideas, particularly on liberty, the title of his most famous work, though he did spend three years as Liberal MP for Westminster in 1865–68, driving the Tories mad by describing them as "the stupid party." He wanted to get rid of all barriers to self-fulfillment. He became associated with a minimal "night-watchman state" (he never used this expression, but it stuck), where the only justification for the state interfering in people's lives was to prevent them from doing harm to others. In Mill's view, economic and intellectual freedom was not just good for an individual; you also got a richer, fairer society if opinions were constantly tested—established orthodoxies and frail egos be damned.

These ideas found political champions in a country where rural hierarchy was being swept away by industry, science, and the cult of efficiency. In the half century after Waterloo, a succession of liberal-minded governments dismantled most of the components of the Old Corruption—from the East India Company to the Corn Laws to rotten boroughs.[19] Nobody personified this more than Gladstone. The erstwhile Tory broke with his party over free trade, apologized for his father (describing slavery as "by far the foulest crime that taints the history of mankind in any Christian or pagan country"), and turned into one of the more radical social reformers ever to hold office.[20] Queen Victoria described him as a "half-mad firebrand." As chancellor of the exchequer and four-time prime minister, he masterminded an astonishing range of reforms, opening civil service jobs and Oxbridge fellowships to competition, inventing the modern company, abolishing the purchase of army commissions, and much else besides.

"The People's William" was a passionate defender of the poor and advocate of equal opportunity—publicly, he gave ever more men the vote and introduced a national curriculum; privately, he devoted his evenings to trying to rescue fallen women. If a figure of 100 represented the energy of an ordinary man, one of his sidekicks observed, and 200 that of an exceptional man, William Gladstone's energy was at least 1,000.[21] But this was no Big Government Liberal. First, Gladstone believed that the poor had a duty to help themselves out of poverty. An unbending moralist, he was driven to prosecute sin wherever he found it, whether in the mansions or the slums.[22] Second, he believed in having as small a government as possible. He wanted health care to be provided by charitable hospitals: "Nothing should be done by the state which can be better done or as well done by voluntary effort." Every administrative expense was scrutinized carefully (saving "what are meant by candle ends and cheeseparing in the cause of the country") and accounted for publicly.[23] He regarded income tax as immoral, because it tempted statesmen to extravagance and taxpayers to fraudulent evasions, and did everything he could to abolish it, so money could "fructify in the pockets of the people." He reduced it from seven pennies to four pennies in the pound.

Mill and Gladstone took the flabby nation-state that won Waterloo and modernized it, opening up government to many of the concepts that politicians mouth nowadays but seldom enact: efficiency, competition, honesty, meritocracy, good government. Inspired by Victorian Britain's success, plenty of other countries copied small government liberalism—not least the United States, where distrust of government ran high. Abraham Lincoln, who was born in the same year as Gladstone, championed the same mixture of limited government and social reform, though with bloodier consequences because the agrarian ruling class in America took the form of southern slaveholders.

In the second half of the nineteenth century, a new division opened up in liberalism prompted by the very urbanization and industrialization that it was helping to speed along. Although Gladstone remained committed to small government, Mill and his wife, Harriet Taylor, moved to what we would now call the left. The same savage inequality that stirred novelists like Charles Dickens and Elizabeth Gaskell and philosophers like Karl Marx (who had fled to liberal London from oppressive Prussia) also made Mill ask questions. What good was all this liberty to a child who had no education? Freedom of opportunity did not mean much if a rich dunce from Eton was better equipped than an uneducated genius. With each edition, Mill's *Political Economy* worried more about social reform and less about liberty.[24] He was not alone in this journey. The poor were no longer layabouts; they were victims.

By the time Mill died in 1873, pressure on government to do more was mounting. Though it would be nice to report that the elite's sudden sympathy for the "left out millions" was caused purely by guilt or Christian piety, something else was also at work: the spur of competition. In 1871, Bismarck finally united Germany: Prussia boasted the best army in Europe—but also good schools and a pension system. Germany followed the creed of G. W. Hegel, who famously described the state as the "march of God on earth" and believed that government's job is to promote the general good over selfish businesspeople. By 1900, Germany scared the hell out of the British establishment in much the same way that China bedevils Washington, DC, today. With Britain's share of world trade dropping and its soldiers humiliated by Boer farmers in South Africa, the establishment decided it needed a healthier, more productive workforce. The School Medical Service, the prototype of the National Health Service, was founded in 1907, followed by pensions (1908), a budget against poverty (1909), and national insurance for the sick and unemployed (1911).

This was the beginning of a great swing toward state activism across Europe. Hobbes's central point, that Leviathan's first duty was to protect the citizen from himself, was being reinterpreted. Security no longer just meant physical security: it encompassed a broader range of services and rights. In the twentieth century, this river split into two streams. For too much of the world, including Russia and China, it resulted in Communism. Ironically, Marx himself was not much exercised by the size of the state or its political forms. For him, it was all about the class struggle, with the state being one of the instruments of oppression of the masses.

Get rid of classes, he argued, and the state would wither away into administration. Sadly, the Marxist regimes who claimed to follow him did not see it that way. Lenin, Stalin, and Mao all maintained that the only way for people to be equal was for the state to administer (and indeed own) everything. Marx's "administration of things" turned into treating people as things. From not mattering at all, the state became everything. And so the totalitarian nightmare began—with Fascism as a counterpoint. For Hitler, the state was the embodiment of race, not class, but the impact on individual liberty was the same.

Most of the West, however, did not veer off in that direction. It hung on to democracy and liberty, just adding a lot more equality and rights. Its answer was the welfare state, with the function of government being to provide an "enforced minimum for a civilized life." The history of the Western state over the past century has been of government gradually increasing that enforced minimum—whether driven by the new parties of the left, centrist Christian Democrats, or One Nation Conservatives. The Western welfare state that sets the pattern for our lives today owes far more to Beatrice Webb than to Karl Marx.

THE WELFARE STATE

If Mill and Gladstone must have been irritatingly smug as children, Beatrice Potter, as she was christened, was insufferable. "The cleverest member of one of the cleverest families in the cleverest class of the cleverest nation in the world," as she put it, was born in 1858—to a rich businessman and an intellectual mother who ran a salon for laissez-faire economists.[25] Like Mill, Beatrice began as a small government liberal but quickly moved to the left, a movement that speeded up when she met Sidney Webb in 1890 and decided that, although he looked a bit like "a tadpole," he was her soul mate. Together the Webbs argued that the only answer to most problems was more government—"collective ownership wherever practicable; collective regulation everywhere else; collective provision according to need of all the impotent and sufferers, and collective taxation in proportion to wealth, especially surplus wealth."[26]

The Webbs never produced a book to match Hobbes's *Leviathan* or Mill's *On Liberty*, but they did create an intellectual movement, founding the Fabian Society, the London School of Economics, and the *New Statesman*. Their aim was to change the collective mind by a policy of "permeation." Some of their views were dotty—they endorsed eugenics, teetotalism, and Soviet-olatory. But they were enormously influential—not just within Britain's Labor Party but across the Western world. Many of the leading intellectuals of the time, from R. H. Tawney, the author of *Equality*, to Herbert Croly, the founder of the *New Republic*, sat at their feet. Even more traditional sorts such as Winston Churchill embraced their idea of a "national minimum."

Modern economics headed down the same path. John Maynard Keynes always saw himself as a liberal who disliked the

Webbs' puritanism as much as their collectivism and believed that the state should never consume more than about a quarter of GDP.[27] But his book *The General Theory of Employment, Interest and Money* (1936) ripped apart laissez-faire capitalism. Left to itself, the market would not prove self-correcting, Keynes showed. Capitalism might be wonderful, but it was also self-destructive and needed to be saved from itself. The invisible hand of the market had to be guided by the visible hand of the state. During crises, government had to step in to boost demand by spending public money, particularly on infrastructure and unemployment pay.

Events sealed the argument in favor of bigger government. The Wall Street crash proved that unrestrained capitalism might end in ruin, while Franklin Roosevelt's New Deal between 1933 and 1939 proved that government spending worked. As Felix Frankfurter, the future great Supreme Court justice, noted in 1936, all the cleverest people went to Washington. "They find satisfaction in work which aims at the public good and which presents problems that challenge the best ability and course of man."[28] Then the Second World War—at least from the victors' point of view—proved the argument all over again. Top rates of tax rose above 90 percent in both the United States and Britain, every industry became a state-directed one, and the great theme of the home front—the need for common sacrifices as everyone "dug together"—transmuted in peacetime to an immediate call for common protections. Across the West, things that had been seen as privileges became entitlements for all—free secondary education, unemployment pay, and, in Western Europe at least, free health care.

The reconstruction of Europe was directed by newly self-confident elites—by the products of the *grandes écoles* in France, advocates of social markets in Germany, idealistic socialists in Sweden, and, above all, the architects of the Common Market. The French and the Germans basked in their economic success. In many cases, the builders were Christian Democratic parties: conservatism and Catholicism were now fully behind the state. Even on the European left, the watchword was centrism. The Swedes built the "people's home," but, to begin with, they made sure that business was included.

This was a golden age to be a European civil servant—not just because pay was in line with the private sector but because of what you were trusted to do. The philosophy was, as one Labor intellectual put it, that "the gentleman in Whitehall really does know better what is good for people than the people know themselves." And, with the same principle applying to gentlemen (and women) in Bonn, Paris, and Rome, bureaucracies began to sprawl. In 1955, Cyril Northcote Parkinson invented Parkinson's Law—that work expands to fill the time available for its completion—partly on the basis of observing the way that Britain's Colonial Office gradually grew as the Empire shrunk, with the maximum number of staff occurring when it was folded into the Foreign Office due to the lack of colonies left to administer.[29]

The United States never went in for nationalizing industries as they did in Europe. But the direction of travel was the same. The top rate of tax stayed above 90 percent till 1964. America wanted to reward the returning GIs with places at university and cheap houses. It needed freeways to move goods around the country and a military-industrial complex to fend off Communism. From the White House, the Republican Dwight Eisenhower and the Democrat John Kennedy both expanded government pragmatically. "The ideological debates of the past began to give way to a new agreement on the practicalities of managing a modern economy," Arthur Schlesinger Jr., Kennedy's house historian, observed. "There thus developed in the Kennedy years a national accord on economic policy—a new consensus which gave hope of harnessing government, business and Labor in rational partnership for a steadily expanding American economy."[30]

There was a clubbish feel to all this. In Britain, a Magic Circle of Old Etonians gathered around Bobbety Salisbury, the grandson of the great nineteenth-century prime minister, to choose the next Conservative Party leader (and thus usually prime minister). Eisenhower and Kennedy called on a cohort of Waspy "wise men," typified by W. Averell Harriman, the son of a railway tycoon. Whenever a problem emerged, the East Coast establishment called in the best and the brightest from academia, business, or the civil service—whether it was asking Keynes and Harry Dexter White to design the IMF and the World Bank, "Mac" Bundy to mastermind foreign policy, or Bob McNamara and his "whiz kids" to fix the faltering Vietnam War. In France, Charles de Gaulle ruled more regally, but by embracing the *Ecole Nationale d'Administration* he developed an elite corps unmatched in Europe. Since 1974, every French president except François Mitterrand and Nicolas Sarkozy has been an *enarque*.

The mid-1960s saw the moment when Western government was at its most self-confident. Lyndon Johnson promised to create a "Great Society" (a phrase stolen from the title of a book by Graham Wallas, a close friend of the Webbs) by reducing poverty, banning racial discrimination, and improving educational opportunities. One reason why Hubert Humphrey, the Democratic left's champion, wanted to stay in Vietnam was that he thought that it too could benefit from having a Great Society of its own.[31] With America's economy doubling in size every decade, economists agonized about "fiscal drag"—that if they didn't spend money fast enough fiscal surpluses would eventually act as a deflationary break on economic growth—so they tried to spend even more.[32] "I'm sick of all the people who talk about the things we can't do," LBJ said in March 1964, seven months before his landslide win over the libertarian Republican candidate Barry Goldwater: "Hell, we're the richest country in the world, the most powerful. We can do it all."

CHAPTER TWO

THE DECLINE OF THE WEST

n his memoirs Edward Gibbon pinpoints the moment when he decided to write *The History of the Decline and Fall of the Roman Empire*, the great work that took up twelve years of his life. "It was at Rome, on 15 October 1764, as I sat musing amidst the ruins of the Capitol, while the barefoot friars were singing vespers in the Temple of Jupiter, that the idea of writing the decline and fall of the city first started in my mind."[1] Will someone one day look at Washington, DC, London, or Brussels the same way? They still look fairly well kept up at the moment—too sleek some would say. But most losses of power tend to be gradual, spasmodic, and, for a period at least, reversible.

Gibbon began his tale of Rome's decline in AD 98, the year a militaristic Spaniard, Trajan, became emperor. Trajan went on to expand the empire to its greatest size, at least in terms of territory. Its decline did not really become obvious until a century later and the fall did not begin till 376, when the empire's borders collapsed. And throughout Gibbon's story there were moments when Rome seemed capable of rescuing itself under good emperors such as Diocletian (who ruled from 284–305). The same applies to the story we have already told about China: the Middle Kingdom possibly reached its peak with Admiral Zheng He's last voyage in 1433, but it had plenty of chances to reverse its decline after that.

So when we date the decline of the Western state to the 1960s, this comes with the fervent hope that it can still rebound. We have the technology, the power, and the competitive threat to prompt a new beginning. But it won't be easy. Since the 1960s, government in the West has tried a variety of cures—from a genuine attempt at revolution under Ronald Reagan and Margaret Thatcher to the quack cures of the modern populists. It has had moments of Trajan-like triumph—especially the fall of the Berlin Wall. But even when it was lecturing the rest of the world about the inevitability of globalization in the 1990s, the Western state never regained the confidence at home it had in the 1960s. The public sector has lagged ever further behind the private sector. And gradually the East has begun to catch up. As Gibbon also mused, "Instead of inquiring why the Roman empire was destroyed, we should rather be surprised that it had subsisted so long."

THE LEFT'S OVERREACH

The starting point was overreach. Britain provides an example. Back in 1914, "a sensible, law-abiding Englishman could pass through life and hardly notice the existence of the state, beyond the post office and the policeman," as the historian A. J. P. Taylor put it. By the mid-1970s a Briton couldn't move without bumping into the state. Leviathan was promising to deliver fairness, equality, happiness, the end of racism, and free opera for the masses. Almost half of Britain's national income was devoted to public spending and nearly a third of the labor force worked in the public sector.[2] There were so many benefits that the Department of Health and Social Security produced a leaflet that listed all the other leaflets. But it still wasn't working. The expert "gentlemen from Whitehall" turned on the Keynesian taps only to produce stagflation; they crammed poor schoolchildren into "comprehensive" schools that failed and their parents into concrete high-rises that became aerial slums. Trade unions turned striking into a regular ritual and, by the mid-1970s, the country seemed to be spiraling out of control. "Goodbye Britain, it was nice knowing you," crowed the *Wall Street Journal* in 1976, as the Labor government went, begging bowl in hand, to the IMF.[3]

Britain was an extreme case, but no part of the West looked healthy. This was the era of oil shocks and dystopian films such as *Death Wish* (1974) and *Taxi Driver* (1976). For America, the Vietnam War ended in humiliation. Richard Nixon tried a variety of ways of kicking the economy back to life, including quitting the gold standard and freezing prices, but nothing worked. He worried in private that the United States had "become subject to the decadence which eventually destroys a civilization"—before becoming, through Watergate, a symbol of that decadence himself.[4] In Europe, the Baader-Meinhof Group and the Red Brigade terrorists were on the rampage. Germany experienced negative growth for the first time since the Second World War, while France was forced into a humiliating devaluation of the franc. In Sweden, taxes rose to prohibitive rates: in 1976, Astrid Lindgren, the creator of Pippi Longstocking, received a tax bill for 102 percent of her income and produced a fairy tale about a writer, Pomperipossa, who gave up producing books for a carefree life on the dole.

There were plenty of people in trade union halls and even more on university campuses who thought that the answer was to head even further to the left. But that strategy ran up against common sense. Why would more government be the solution to problems that more government had failed to solve? And what about the increasingly obvious disaster in the Communist world? "Government cannot solve our problems," Jimmy Carter, the Democratic president, told the American people. "It cannot eliminate poverty or provide a bountiful economy, or reduce inflation, or save our cities, or cure illiteracy, or provide energy."[5] Middle-class people began to turn against the Great Society. They complained that their taxes were being wasted on the underclass while the state was failing to provide law and order. There was racism, snobbery, and absurd amounts of hypocrisy in this: the leafy suburbs never complained about the public subsidies for their mortgages, their children's free university education, or the local theater group. But bourgeois frustration with Leviathan rekindled the intellectual heirs of John Stuart Mill.

The free-market right had never actually disappeared. Friedrich Hayek's *The Road to Serfdom*, published in wartime London in 1944, was a bestseller on both sides of the Atlantic. In 1947, the so-called Austrian school of free marketeers met at Mont Pelerin to lay the foundations of a reverse Fabian Society. In the 1950s, the movement's center of gravity moved from Europe to Hayek's new home, the University of Chicago, where a group of economists explained how bureaucrats, even if they were well intentioned (which they often weren't), could do more harm than good.

The loudest voice was the American economist Milton Friedman. Small, wiry, intense, Friedman attacked just about everything that the center-left held dear: foreign aid was siphoned off by dictators, rent controls reduced the supply of housing, public spending on universities forced poor people to subsidize rich ones. He was like a naughty schoolboy frightening old ladies with firecrackers. Big government would always mess things up: "If you put the federal government in charge of the Sahara Desert, in five years there'd be a shortage of sand," Friedman said. He even broke with the standard reformers' line that we must make government more efficient on the grounds that a better government would only improve Leviathan's ability to rob the people.

Like Hayek, Friedman hated being called a conservative. He saw himself squarely in the tradition of John Stuart Mill and Jeremy Bentham. In 1964, his vision of a night-watchman state was embraced by the Republican presidential nominee: "I have little interest in streamlining government or in making it more efficient, for I mean to reduce its size," Barry Goldwater told the electorate. "I do not undertake to promote welfare, for I propose to extend freedom." Running against Lyndon Johnson and the Great Society, Goldwater won just six states. But as the Great Society collapsed, Friedman found two better champions. In 1979, Margaret Thatcher swept into Downing Street, with Hayek in her handbag, and a year later Ronald Reagan won the White House. Then the pair began, in Thatcher's words, "a world-wide revolt against big government, excessive taxation and bureaucracy."[6]

Reagan relished a battle with Leviathan, whether it came in the shape of the Soviet Union or the air traffic controllers union. But Thatcher was bolder in reforming government, partly because Britain was in so much worse shape than America and partly because she didn't have Reagan's luxury of running eye-watering deficits.[7] It required nerves of iron to keep pushing ahead in the early 1980s even as her reforms devastated parts of the British economy—and, electorally, she was probably only saved by the 1982 Falklands war. Lucky or not, she succeeded in changing the country's direction. The number of employee days lost to strikes fell from 29.5 million in 1979 to under 2 million in 1986. The top rate of tax fell from 98 percent in 1979 to 40 percent.[8] And in a wave of "privatisations" she set free forty-six state-owned companies, including British Gas and British Telecom, as well as selling council houses to their occupants.[9]

The fall of the Berlin Wall in 1989 created a mood of euphoria across the West. The Anglo-American hymn of deregulation, globalization, and privatization rang out across the world. Bill Clinton declared the age of big government to be over, and Tony Blair ditched "Clause Four," which had committed the Labor Party to nationalization. The likes of Renault and Lufthansa were privatized. New York and London became the symbols of the new era: cosmopolitan, lively, and revitalized by the power of global finance. An army of Ivy League consultants—many of them little more than teenagers—traveled to Moscow to sell off eighteen thousand state companies. Further east, India and China both experimented with market reforms and lifted more than a billion people out of poverty.

For a while in the 1990s it looked as if trade, technology, and finance would unite the world—and shrink the state. Just as powerful central bankers tamed inflation, people speculated that the bond markets would bind Leviathan. Even socialist Sweden eventually buckled. Public spending there was chopped from 67 percent of GDP in 1993 to 49 percent of GDP, while the top marginal tax rate was cut by 27 percentage points.

FAREWELL, WASHINGTON CONSENSUS

So was this another revolution? Not really—or certainly not enough of one. At home, Thatcher and Reagan did more to change the debate about the state than they changed the state itself (possibly making reform more difficult). Abroad, the Washington consensus overreached.

Privatization aside, neither Reagan nor Thatcher succeeded in reinventing the public sector for the global age they trumpeted. Reagan's victory over the air traffic controllers did not change much: a study in 1996 found that air traffic controllers reported through sixteen layers of decision-makers (with the number rising to sixty layers when it came to policy and budget questions).[10] In her eleven momentous years in office, Thatcher succeeded in reducing social spending from 22.9 percent of GDP in 1979 to 22.2 percent in 1990. With his spending cuts stymied by Congress, Reagan financed his tax cuts mainly through debt. And even though spending did dip under Clinton and Blair, it soon grew again—especially under George W. Bush and Gordon Brown. By 2010, with the British economy struggling, the state's share of GDP was back above 50 percent.[11]

Moreover, the official numbers masked the other ways in which Leviathan was growing. Even if both Britain and America limited the number of civil servants they employed directly, the number of contractors ballooned, creating a series of opaque shadow states. One of the things we learned during the Snowden affair was that half a million private contractors had top security clearances at the National Security Agency. And regulation expanded, especially in America, where the Federal Register, by Niall Ferguson's calculation, grew two and a half times faster than the economy. In the first decade of the twenty-first century it increased at a pace of 73,000 pages a year, compared with an average of 11,000 pages a year in the 1950s.[12] Barack Obama's 2010 Affordable Care Act was 2,700 pages long, including a 28-word definition of a "high school" and 140,000 codes for ailments.[13] By contrast, FDR's landmark welfare legislation in the 1930s had been just 30 pages long. At the state level, America created a "License Raj," bossing around occupations that posed no plausible threat to health or safety such as florists, tour guides, frozen-dessert sellers, second-hand booksellers, and interior designers.[14]

Politics in the Anglo-Saxon world became an impasse. The left focused on expanding government by stealthy regulation, and the right focused on cutting taxes. When Reagan's successor, George H. W. Bush, broke his "no new taxes pledge" in 1990 (largely because of Reagan's deficits), activists at the Heritage Foundation paraded his effigy around. On the American right, there was only one answer to any debate about the state: less government. Shutting down the executive became a badge of honor. "By the time we finish this poker game," Tom DeLay, a rising congressman from Texas, said of one stand-off in 1994, "there may not be a federal government left, which would suit me fine."[15] No matter that the same Republicans were happy to spend Uncle Sam's money on boondoggles for the rich and never dared touch entitlements; they would zealously stick to their "Reaganite" promise not to raise taxes. Serious debate about how to reform government ground to a halt.

If the gap between rhetoric and reality was starkest in America and Britain, continental Europe soon noticed what was happening and gave up on the rhetoric altogether. By the first decade of the twenty-first century it was still a *faux pas* to proclaim in public that you were for big government, even in Brussels. But the Common Agricultural Policy kept harvesting 40 percent of the European Union's budget, and a cavalcade of courtesy limousines queued up outside the Commission's headquarters, the Berlaymont, to whisk eurocrats to another important lunch. Individual states went further: France reduced the working week from thirty-nine hours to thirty-five in 2000. Twelve years later, it briefly added a 75 percent tax on top earners.

By then it was safe to say that the free market had been repulsed. Milton Friedman had long before admitted defeat, writing in 2004, a couple of years before he died: "After World War II, opinion was socialist while practice was free market; currently, opinion is free market while practice is heavily socialist. We have largely won the battle of ideas (though no such battle is ever won permanently). We have succeeded in stalling the progress of socialism, but we have not succeeded in reversing its course."[16]

At the same time the spell of globalization had begun to fade. Economically, it was still achieving miracles in the emerging world, dragging a billion people out of absolute poverty, but politically it looked tarnished. Despite all that help from Harvard, Russia botched its privatization program, enriching kleptocrats and giving a former KGB colonel called Vladimir Putin a chance to sneak into power at the end of 1999. The invasion of Iraq in 2003 that was supposed to launch a wave of democratization in the Middle East became a bloody quagmire. In Europe freedom of movement created a backlash, particularly in Britain, while the single currency destabilized the Mediterranean economies.

Then came the financial crash of 2007–8. How could neoliberals continue to preach the gospel of free markets when Wall Street had unleashed havoc? And what happened to the promise that even if globalization made the rich richer, all boats would still rise? Incomes had barely budged, making fairness the great political issue. In Europe, the euro crisis added another twist. If you were German, why were you bailing out the spendthrifts in the South? If you were Greek, why were the Germans imposing rule by technocrat? And if you were British, what was the point of being part of this mess? All this paved the way for populism. But this increasingly fractious West now faced a rival, whose rise had been accelerated by the same globalization that the West preached.

THE ASIAN RENAISSANCE

The revival of government in the East began in a tiny island-state that had once been a cog in the British imperial machine, and its presiding genius was a Jane Austen–reading former Fabian whom one British foreign secretary described as "the best bloody Englishman east of Suez."[17] Lee Kuan Yew got the best school certificate results in Singapore in 1940, when the country was a British colony, won a scholarship to Cambridge, and graduated with a double starred first in law (one of the few students to match his grades later became his wife). He subscribed to the main tenets of socialism and even campaigned for the Labor Party. But after masterminding Singapore's independence in 1959, he moved to the right.

Lee challenged what he saw as the main flaw of the postwar welfare state: its overgenerosity. He blamed Britain's decline on its "all you can eat buffet" of benefits. The state Lee created was small in terms of the percentage of GDP. It forced people to provide for their long-term welfare through a system of self-insurance: Singaporeans had to pay a fifth of their salaries into the Central Provident Fund, with the state contributing another 17 percent, in order to pay for their housing, pensions, health care, and their children's university education. Like Gladstone, he preferred self-reliance: the care for the old and weak was the duty of the family, supplemented by the state, rather than of the state alone. Lee fumed about the degenerate "Western" idea of children stuffing parents into government-financed care homes. He also made people pay a small fee to visit doctors, to discourage the overuse that happens with the NHS.

But at the same time Lee also challenged the right-wing assumption that government was the problem rather than the solution; he had no truck with the childish antics of Reagan's heirs. He focused on bringing talent to the public sector, creating a modern mandarin class trained not in the classics but in the sciences.[18]

Bright students were given generous scholarships to study abroad providing they then worked for the state. Even today, half the students who go from Raffles Institution, Singapore's most elite school, to foreign universities sign bonds to pay back their college education through government service. Lee paid his bureaucrats well. The pay of junior ministers and permanent secretaries is normally above a million dollars (though their bonuses were cut this year because of Covid).[19] Again breaking with the Western right, he gave these prodigies the task of guiding the economy from manufacturing to services to the knowledge industries.

Lee's new model produced particularly striking results in health care and education. Singapore has the longest life expectancy in the world and the lowest infant mortality rates, thanks to its public health system, but its private hospitals also attract medical tourists from across Asia. Singapore's schools routinely finish top of the league tables in terms of results, but not spending, simply because it rewards good teachers and sacks bad ones. The system is proudly elitist: you can't get a job in teaching unless you've graduated in the top third of your class (a rule that also applies in Finland and South Korea, which also shine in the rankings) and the government will pay for your university education if you agree to teach for around five years after graduating. Examinations and streaming have always been the rule, with bright children picked out early and fast-tracked.

There is an authoritarian edge to all this: Lee's party has ruled since independence. "We decide what is right," he once observed.

"Never mind what the people think." Like Hobbes, Lee believed that a social contract makes demands on citizens as well as rulers. People who engage in littering (or nowadays break social distancing rules) are subjected to fearsome penalties. The result is a very different state from the one that is struggling in the West: elitist where the West is democratic; stingy where the West is (over) generous; focused where the West is sprawling; and proudly interventionist in the economy where the Anglo-Saxon West at least is hands-off.

The Asian tigers have either copied Lee's model directly, or arrived at roughly the same destination independently. They have all invested heavily in education with an emphasis on competition and testing. South Korea has driven growth through national champions like Samsung and Hyundai. Indonesia and the Philippines are basing their health-care systems on Singapore's social insurance system rather than, say, Britain's tax-based system. They all want to make their civil services more elitist, so South Korea has a Senior Civil Service, Thailand has its High Potential Performers, and Malaysia even has something called TalentCorp. Many of them are now much more socially liberal than Lee; it is hard to imagine Singapore producing K-Pop or *Parasite*.

Lee's most important student, though, lies to the north. China knows it lost its global preeminence because its state withered. It is easy to mock its goal of becoming a giant Singapore. Even now, the World Bank ranks China in the sixtieth percentile for government effectiveness and the fortieth percentile for the rule of law.

It has had endless struggles with corruption. In 2014 the authorities discovered one bureaucrat had the entire basement of his twenty-thousand-square-foot house stocked with cash of various denominations that weighed more than a ton. Much of China is still wretchedly poor while the red princes and princesses glide from St. Moritz to the Ivy League, protected by their parents.

But alongside the thuggish dictatorship there is another China: one that studies where government works and where it doesn't; that is recruiting a cadre of highly trained administrators and monitoring them through the Party's Organizational Department. Civil servants have to prove their mettle. If you are running a province, how much has the economy grown? If you run a university, how much have you boosted student enrollment? Whereas America's rulers tend to be lawyers, China's are engineers. When Xi Jinping, who studied chemical engineering at Tsinghua University, joined the Standing Committee of the Politburo in 2007, all nine members were engineers.

Back then, China appeared to be sticking to the idea that it would change its leader every ten years (thus avoiding the autocracy trap of strongmen staying on too long). So when Xi took over from Hu Jintao in 2012, he was expected to stand down in 2022, but he ditched the ten-year rule in 2018. The full impact of that has yet to be seen. Sooner or later, China's growing middle class will *surely* demand more freedoms, as their peers in Taiwan and South Korea have already done. But don't underestimate the Chinese regime. It has held the country together, despite cramming a century's worth of economic changes into just a couple of decades. It is rapidly building a health-care system, and it is well on the way to becoming an educational superpower. As in Singapore, children compete to get into the best nursery schools so that they can get into the best secondary schools so that they can get into the best universities. The Chinese compare the *gaokao* exam for universities to "ten thousand horses crossing a river on a single log."[20]

THE POPULIST CONTAGION

In the past, the West might have responded to the sound of the ten thousand horses with reform of its own. Instead, many countries have taken refuge in populism.

There have always been crowd pleasers in politics: in the *Republic*, Plato worried about the way that democratic societies fall under the spell of demagogues who pander to people's basic instincts. At the end of the last great age of globalization in the nineteenth century, American farmers blamed grain prices on "the money power" of the railways and the bankers. In the 1930s, European populism took on its ugliest form under Hitler and Mussolini. But Britain had Oswald Mosley's black shirts and America had Huey Long, the governor of Louisiana, who promised to make "every man a king." And even established politicians fished in their waters: in 1936, Franklin Roosevelt worked Madison Square Garden into a frenzy by telling the crowd that he relished the hatred of the rich.

During the long postwar boom, populism retreated to a few islands of discontent: France's National Front drew its support from the *pieds-noirs* forced out of Algeria after decolonization and from small shopkeepers who hated paying taxes. As the welfare state began to fray at the edges, Nixon championed "the silent majority" against limousine liberals. Resentment intensified as trade barriers came down and immigration went up. In 2002, the National Front's leader, Jean-Marie Le Pen, finished second in the French presidential election. And, as the private sector soared, the idea of a brash businessman who could fix everything became more attractive. Ross Perot won one in six votes in the 1992 US election. For all his clownish incompetence, Silvio Berlusconi persuaded Italians he should be their prime minister four times.

Over the past decade, however, populism has come into its own by successfully exploiting global forces that mainstream politicians failed to understand. The most obvious was resentment against the global elite, particularly its habit of "failing upwards," like the bankers who were bailed out after wrecking the global economy. Nationalism also reappeared. The elites preached the world was flat. Many others wanted the red meat of flags and anthems rather than the tasteless acronym soup of EU, NAFTA, the WTO, and (if you are a Scottish nationalist) the UK. Ironically, the final force is technology: the populists have been more at home in the world of tweets, Facebook pages, and electronically shared conspiracy theories. Beppe Grillo founded the Five Star Movement, now the largest party in the Italian parliament, on a blog in 2009.

The populist revolution began on the periphery—in Eastern Europe (Viktor Orban in Hungary and the Law and Justice Party in Poland) and the euro-ravaged Mediterranean (Podemos in Spain, Syriza in Greece, and a cluster of parties in Italy).[21] In 2016, the barbarians reached the gates of Rome. First Britain voted for Brexit. "Isle of madness" wrote *Der Spiegel* in despair, and the *New Yorker* displayed a group of bowler-hatted lemmings walking off a cliff. Then Donald Trump beat Hillary Clinton to secure the White House.

Since then, the populist effect has been most striking on the global stage. The West is losing its collective ability to act as a voice for free trade and free minds. Not only has nationalism widened the divide between Europe and America; the West is now led by a man who loathes globalization, wants to quit pretty much every global institution, and disdains the language of Liberty. Trump has confronted China on its trade abuses, but from the perspective of America First. Despite its success, or perhaps because of it, democratic Asia has been shunned; one of the first things Trump did was to extract the United States from the Trans-Pacific Partnership. He has made clear, in his repeated conversations with dictators, that human rights are not a priority.

At home, populists have generally ducked the challenge of reviving the Western state. There are certainly reformers in their midst. The White House contains a small group of deregulators who have cut the number of pages in the *Federal Register* from its record under Obama. Dominic Cummings, Boris Johnson's chief adviser, sees the break with Europe as merely the first stage of bureaucracy-slaying that will eventually create a "meritocratic technopolis." But these would-be reformers look outnumbered. For every Brexiteer who wants to build a Singapore-on-Thames, there are many more who want their newly independent state to protect them against globalization—and Johnson seems intent on building a bigger state for them. In Trumpland, reform has been drowned out in the cacophony of tweets.

It is hard to blame the populists for not redesigning the state; after all, their predecessors singularly failed in that too. The worry is that they are making it worse. Just like Berlusconi before them, Trump and Johnson are undermining the idea that statecraft is a serious business; instead they have treated it as a branch of mass entertainment. Constitutional restraints, independent civil servants, central bankers, judges, and experts of all sorts have been derided. In America, more than a thousand scientists have left the Environmental Protection Agency and the Department of Agriculture.[22] In Britain, Cummings's war against "the blob" as he likes to call it has extended to the BBC, the universities, the quangos, the law courts, and the civil service. And truth has often been one of their victims. From the moment that he boasted that the crowds at his inauguration were larger than anybody else's, when they manifestly were not, Trump has dealt in fibs, big and small. When a member of the public asked Johnson during a televised election debate in 2019 whether he valued truth, the audience burst into laughter.

In some ways it is as if we are once again gazing at the fresco of Bad Government in Siena—with Justice bound and the characters of Cruelty, Deceit, Fraud, Fury, and Division looking on. It was into this world that the Coronavirus appeared at the beginning of 2020. We will resume that story in Chapter Four. But first we want to dissect what has actually gone wrong with Leviathan in the West.

CHAPTER THREE

THE OVERLOADED STATE

Writing to his wife in May 1942, Evelyn Waugh recounted a true story of military derring-do. A British commando unit, training in Scotland, offered to blow up an old tree stump on Lord Glasgow's estate, promising him that they could dynamite the tree so that it "falls on a sixpence." After a boozy lunch, the officers and Lord Glasgow all trooped down to witness the explosion. But instead of falling on a sixpence, the tree stump rose fifty feet in the air, taking with it half an acre of soil and a beloved plantation of young trees. A tearful Lord Glasgow fled to his castle, only to discover that every pane of glass had been shattered. He then ran to his lavatory to hide his emotions, but when the distressed peer pulled the plug out of his washbasin "the entire ceiling, loosened by the explosion, fell on his head."

Our plight is similar to Lord Glasgow's—only worse. The populists promised us that they could dynamite the establishment but still leave prosperity and security intact. In fact, they have made the crumbling Western castle's problems worse. Without Covid, it might have taken time to reveal just how dilapidated it had become, especially in comparison with the gleaming new fort being built in Asia. Now, Covid has appeared like a hurricane, ripping the entire roof off. We can blame the populists and the pandemic as much as we want, but even if Donald Trump had stayed in light entertainment and the virus had never left Wuhan, a reckoning was coming. Western government has been crumbling for decades, overloaded with obligations, undersupplied with talent, and picked apart by special interests.

Here are the seven main flaws. They vary from capital to capital—Copenhagen is in better shape than Washington, DC, or Rome. But they explain why the West has found the pandemic so hard to deal with.

1. THE OUTDATED STATE

It's hard to say whether the machinery of Western government has suffered from *absolute* decline. For all its sprawl, some parts of government are far more efficient than they were in the past. You can get your driver's license renewed with a few keystrokes. New

York is clearly safer today than it was in the 1980s. But there is little doubt that Western government is suffering from *relative* decline compared with the private sector: indeed, it now often lives in a different century.

Try to think of any big company or industry that has not changed shape dramatically over the past few decades. The five most valuable companies in America are Microsoft (founded in 1975), Apple (1976), Amazon (1994), Alphabet (1998), and Facebook (2004). Farming is a sideshow. Yet farming absorbs two fifths of the European Union's budget and the US Department of Agriculture still employs 100,000 people spread out across 4,500 locations. America's structure is dictated by a Constitution that was designed for a country of thirteen states and four million people. The Founding Fathers had no plans to bring either North Dakota or California into their union, nor could they have imagined the ramifications of those two states having the same number of votes in the Senate despite one state having fifty-two times more people. In Europe, some constitutions have been redesigned more recently, thanks to two world wars, but many ancient oddities survive. Belgium has three different languages and five parliaments; unsurprisingly, it often ends up without a government. Britain still has some hereditary peers in its upper chamber, while France's presidency still encourages its occupant to behave as if they were the Napoleonic master of all of Europe.

Technology is a weak spot for the public sector across the West: look at the mess most governments, including Germany's, have made of their Covid-tracking apps. But the lapses in America's public sector seem particularly striking, given its lead in technology. America's federal agencies spend $90 billion a year on technology, but despite some obvious centers of excellence, such as espionage, the overall picture is dire.[1] The core job of America's Social Security system—processing disability and retirement claims—relies on sixty million lines of code written in a computer language that was created in 1959.[2] Until 2019, the division of the Air Force that oversees the nuclear missile system used a 1970s floppy disk system, hardly a comforting fact for those of us who want to avoid nuclear incineration. This is perhaps not surprising if you look at the age profile of the people tech departments are employing: according to Max Stier of the Partnership for Public Service, there are five times as many people working in government IT who are above sixty than below thirty.[3]

One reason why the state seems so ancient is that it does not learn. In the private sector you have no choice: if a competitor anywhere in the world comes up with a better product or service, you respond or go out of business. City mayors have been receptive to ideas: look at the way that "Boris bikes" have spread around the world.[4] But central governments are usually blinkered: it is the devil's work to get British educational bureaucrats to learn from schools in Singapore or Finland. In higher education, German and French universities have watched talent (and students) seep away to the Anglo-Saxon world for decades, without changing their habits. One of the most disappointing things about the EU is that it concentrates on regulating its member states rather than helping them to learn from each other.

2. THE OVERSTRETCHED STATE

It is hard to learn when you're crushed under a burden of obligations you can never meet. In the private sector, institutions have flourished by doing a few things very well: Apple provides the design and brains for its phones, not the cases and chips. The public sector is still in the age where the Ford Motor Company used to own the fields on which grazed the sheep whose wool went into its seat covers. Apart from selling off a few nationalized companies, the Western state has not shed any responsibilities over the past century—and has accumulated many more. It is supposed to deliver an astonishing range of services—from the essential (education for the young and pensions for the old) to the bizarre (advice on how to indulge in "chem sex" for those so inclined) to the pointless (rules on hair-braiding and interior design).

The welfare state was supposed to be a safety net for the poor and unfortunate: instead it has become a collection of overstuffed cushions for the plump bottoms of the middle classes. Higher education and housing are two areas where governments often intervene to make rich people richer. Most European countries give free university education to students who are richer than the average citizen (and then those students pocket most of the gains from their education). In America, it is mainly the well-off who read the

IRS's ninety-page booklet that explain the fifteen different tax incentives for higher education. For all the blame that unfettered capitalism took for the subprime mess, there were few more culpable organizations than Fannie Mae and Freddie Mac, the giant government agencies that pumped money into the housing market. Rent control has increasingly become a perk for the privileged, whether it be in Stockholm or New York. More money goes into subsidizing mortgages than social housing for the poor.

This fits a pattern. No matter how much Americans moan about their government, more than half the households in the country get some kind of handout. About 120 million Americans claim benefits from two or more programs.[5] And that does not count all the nonfinancial demands that the prosperous put on the state.

Whenever crime rises or a cache of immigrants arrives, the cry goes up: "Something must be done." Rules are added—and then someone has to administer them. Whatever its merits, the European Union has been a great complexity-adder, wrapping a new (and often confusing) layer of red tape around each of its countries. Again, the reason behind each strand is usually fear that something somewhere will go wrong.

3. THE OPAQUE STATE

Indeed, government is by its nature a matter of rules and regulations: the term "bureaucracy" means the government of "offices" rather than arbitrary individuals. Many of these rules have admirable aims: better health care, cleaner air, less discrimination against minorities. But the complexity that these regulations engender also imposes a considerable cost.

In some cases, they merely waste time. Do European consumers really need labels on cans of salmon warning them that they "may contain fish," or rules to limit the power of vacuum cleaners? In others, they skewer markets: America's housing crunch was partly caused by microregulations designed to encourage home ownership. Europe's inability to create small companies is largely explained by regulations. In Italy, you need to be a big company (or well connected) to find your way through the Dario Fo world of permissions and licenses. But America is catching up. In the US tax code, there are forty-two different definitions of a small business. To open a restaurant in New York, you need to deal with eleven different city agencies. And then there are all the occupational licenses that force aspiring barbers in Texas to study coiffure for more than a year and aspiring wig makers in Arkansas to take written exams.[6]

The main victim of all this red tape is government itself. One reason why infrastructure projects take so long is that officials have to jump through so many hoops. Italians have been staring at plans for a bridge linking the mainland to Sicily since the 1990s. Berlin's airport is another tawdry tale of bureaucratic bungling and engineering incompetence. One day, presumably after airplanes have become extinct, Heathrow may have a third runway. During the Great Depression it took four years to build the Golden

Gate Bridge. Today, as Philip Howard of Common Good has doc-
umented, bigger highway projects take a decade just to clear the
various bureaucratic hurdles before workmen can actually get
to work. When the New York Port Authority decided to upgrade
the Bayonne Bridge, which arches spectacularly between Staten
Island and New Jersey, so that new super tankers could glide un-
derneath it, it had to get forty-seven approvals from nineteen different government departments, a process that took from 2009 to
mid-2013.

This ties back into the failure to learn: officials are so busy
with paperwork they seldom get a chance to see what works elsewhere. The state operates in a murky, Dickensian world, reminis-
cent of the opening of *Bleak House* ("Fog everywhere. Fog up the
river . . . fog down the river . . . Fog creeping into the cabooses of
collier-brigs . . . fog drooping on the gunwales of barges and small
boats . . . Fog in the eyes and throats of ancient Greenwich pensioners . . . fog in the stem and bowl of the afternoon pipe of the
warful skipper"). Fog inserts itself into every piece of legislation
and every document. It confuses everybody who has anything to
do with the state. It hangs thick over every department. It distorts
every relationship between government and citizen.

4. THE CAPTURED STATE

This fog leaves government at the mercy of interest groups. Indeed
complexity and capture facilitate each other. The explanation for

why lobbies thrive in democracies was deduced brilliantly back in 1965 by the economist Mancur Olson. In *The Logic of Collective Action* he argued that narrow constituencies pursuing specific goals are much more potent than broad ones pursuing general goals. The latter are plagued by free riders who want to enjoy the benefits of political action without incurring the costs.[7] In narrow groups, individuals can see the rewards much more clearly, and they are more likely to pay up and get organized. So Leviathan has been stealthily picked to pieces by scores of narrow interest groups seeking particular exemptions and obscure boondoggles, normally hidden from public view.

The interest groups come in two main varieties: insiders, especially the public sector unions, who see the state as their bread and butter; and outsiders, especially companies and industry lobbies, partisan groups, and demographic groups, that try to divert the state's attention to their cause. Both groups are barriers to government reform.

The leading insiders are the public sector unions. Unlike their weaker private sector equivalents, they can shut things down without risking their jobs. For a month before Christmas 2019, Paris was in chaos because the metro and train drivers were on strike over their pensions. Public sector unions are also unusually good at hiding self-interest, right down to their names: thus the British doctors union is the British Medical Association and the American teachers union is the National Education Association. A few unions have political ties to the right—the alliance between

California's prison guards and local Republicans helped both build jails and pass tough laws to fill them. But the strongest ties are inevitably with the left. In Britain, the trade unions have the biggest say on who becomes Labor leader. In the United States, the teachers unions have a lock on the Democrats' education policy: an old journalist's trick when standing next to someone on the floor of the Democratic convention is simply to ask, "What school do you teach at?"

Protecting jobs, rather than improving services, is the priority. After the killing of George Floyd, people rushed to see why so few American cops have been punished for brutality. They discovered that departments had been persuaded by unions to erase disciplinary records, some after only six months. Derek Chauvin, the man whose knee throttled Floyd, had faced at least seventeen misconduct complaints during his two decades with the Minneapolis police department but had suffered no consequences other than two reprimands. When Chauvin was arrested for murder, his boss, Lieutenant Bob Kroll (who was himself the subject of twenty-nine complaints), responded by accusing the authorities of selling out the police force. Nor are the police that unusual. The federal government fires about one in two hundred of its workers annually compared with about one in thirty workers in the private sector.[8] California has something called "the dance of the lemons" whereby incompetent teachers, rather than being laid off, are bounced from school to school, or left sitting in an office if nobody will take them.[9]

Unions also explain why productivity is so low. They instinctively resist labor-saving devices such as replacing people with machines: introduce automatic doors on Britain's Southern Rail, and the union will insist that you need a guard to watch them. Labor represents 80 percent of the US Postal Service's costs compared with 53 percent at United Parcel Service and 32 percent at FedEx.[10] Pension abuse has been rampant. In Greece, before the euro crisis, many civil servants could retire at fifty on full pensions. In France, SNCF train drivers and Paris Metro staff could slink off at fifty-five. In Rio, the beaches are crammed on weekdays not with girls from Ipanema who make you go "Ah," but with pensioned civil servants in speedos sunning themselves at public expense.

These are the sort of things that drive taxpayers and right-wing politicians mad. But capitalist outsiders are also pretty good at bending Leviathan to their ends. The Common Agricultural Policy doles out more than $65 billion a year, ostensibly to keep dying rural communities alive. In fact, more than 80 percent of the cash goes to the richest fifth.[11] The process is so complicated that almost nobody in the European bureaucracy understands what is happening, let alone voters. But politicians know that farmers will crush them under their tractors if they touch it. Woe betide the French presidential candidate who does not appear at an agricultural fair, clutching a calf and a cheese, and talking about "la France Profonde." And now Central and Eastern Europe oligarchs have got their snouts in the trough, hiding their names behind opaque companies. One company that included the Czech prime minister,

Andrej Babis, collected at least $42 million in subsidies in 2018. In Hungary, Viktor Orban has used the system as a way of building his political machine, while in Slovakia, the top prosecutor has acknowledged the existence of an "agricultural Mafia." When the *New York Times*, which spent a year investigating all this, tried to get statistics from the European Union, it was stonewalled.[12]

America, the home of money politics, is even more prey to interest groups. The US tax code illustrates Olson's law perfectly: every exemption is helping somebody. One reason why Donald Trump is reluctant to reveal his tax returns is that, like most New York real estate developers, he seldom pays tax. The health-care industry has a labyrinth of deductions for health plans. For all the justified protests about offshore tax havens, America's worst tax avoidance scam is at home, particularly in Delaware. Lobby groups cleverly mix greed and fear: you get some money if you support them, but if you don't, your opponent will get a fortune in your party primary. The lobbyists' most seductive gift is not an all-expenses-paid trip but a supply of briefing documents that help an overworked politician to stay on top of the news and cope with the tsunami of words that washes over their desks.

Much of the lobbyists' energy goes into protecting arcane pieces of legislation that matter enormously to one group. The "carried interest" tax break that gives private equity an advantage over other ways of running a company is always due to be scrapped in each tax reform package, but mysteriously never is. Back in 1920, the US Congress, ruminating on the First World War, passed the Merchant Marine Act, decreeing that all goods being transported from one American port to another must be carried by US ships owned by US citizens and operated by US crews. Known as the Jones Act, this piece of protectionism costs the country anywhere from $656 million a year to $9.8 billion.[13] On a bigger scale, the tax break for employer-provided health insurance, which costs Uncle Sam more than $175 billion a year, stemmed from a "temporary" adjustment to Second World War wage controls in order to deal with labor shortages.[14]

In America this extortion is made easier—enforced almost— by its campaign finance laws. Money counts in every democracy but not to the same extent that it does in America. The need to raise cash not only dominates politicians' every waking hour—no wonder they can't think about the future—but also throws them into the arms of special interests, who provide not only money but, in the case of unions, manpower. The Supreme Court justifies all this on the grounds of free expression. Even if that is indeed the case, it comes with a massive democratic cost, leaving American politics in the hands of rich individuals and political dynasties that possess the political machines and name recognition. The country that was born in a revolution against monarchy and inherited privilege is becoming a neofeudal place, with political offices such as ambassadorships sold to the highest bidder.

So everybody with any sense in the West is directing Leviathan to their own ends: teachers, private equity tycoons, prison guards, real estate moguls, French farmers, and East European oligarchs.

And looked at across the West, one group is greedier than all the others combined. Yes, it's your grandmother.

5. THE OLDFARE STATE

When Otto Von Bismarck, at the grand old age of seventy-four, established the first European state pension scheme in 1889, he fixed the retirement age at seventy at a time when average life expectancy was seventy-two. When Franklin Roosevelt established America's Social Security system in 1935, he fixed the retirement age at sixty-five when life expectancy was just sixty-one. Since the foundation of the world's welfare states people have been living much longer but retirement ages have often stayed the same: Americans can still retire at sixty-five but their life expectancy is now seventy-nine. Some countries, particularly in infant-starved Europe, have even *reduced* their retirement ages. In France, when Emmanuel Macron tried to reform the pension system in 2019, the official retirement age was sixty, but there were forty-two different pension regimes, reducing the average still further. Italians talk about "baby pensioners," and abuse of the system is rife, especially if you can claim disability: one fifty-eight-year-old man was caught working as a football referee on weekends despite being officially registered as blind.[15]

In the United States, over 90 percent of all social insurance assistance goes to people aged sixty-five and over.[16] Obviously much of the money is going to retirees who are indeed poor, but many others have lived through some of the most prosperous years in American history. It is also unsustainable. The big three entitlement programs—Social Security, Medicare, and Medicaid—make up almost half of the federal budget.[17] But the government is still not putting enough money into the system: the Social Security fund will run out of money by 2035 and the Medicare fund will run out of money by 2026.

Meanwhile money that is automatically slotted for entitlements cannot be spent elsewhere. The Steuerle-Roeper Fiscal Democracy Index shows that the proportion of discretionary spending has declined from two-thirds in 1962 to a third in 1982 to a quarter in 2014 and is slated to fall below a tenth in 2022.[18] At the same time, public investment in transportation has declined to 1.6 percent of GDP, far less than China's. If you want to know why the average American dam is reaching pensionable age, or why America's trains are now slower on average than they were half a century ago, then ask your grandmother.

6. THE TALENTLESS STATE

The modern economy is increasingly a brains economy: Google and Amazon couldn't exist, let alone thrive, without attracting the brightest workers. This leaves the public sector vulnerable. Even elite jobs in the civil service have lost their cachet. The situation is so dismal that students who study public service would prefer to work in the voluntary sector.

There are some first-rate people in the public sector—brilliant economists, smart diplomats, wily spies. But the bench is far weaker than the private sector. The American economist Tyler Cowen called one of his books *Average Is Over*. It certainly isn't over in the halls of Leviathan. Do an average job and you will still be able to coast to retirement (and, according to one study in America, get paid more than a private sector worker, once you take perks and hours into account).[19] By contrast, do an exemplary job for the Western state and you will never get the rewards you get at Google—or indeed in Asian government. One reason why the Singaporean Civil Service is better run than, say, America's Veterans Administration is that it is led by someone who is paid ten times as much to manage an organization that is half the size.

The demonization of the public sector doesn't help. Trump has repeatedly treated senior civil servants like contestants on *The Apprentice*. Why should the most able people forgo the pay and excitement of the private sector in order to be mocked and insulted? America adds a peculiar disincentive of its own: the raft of political appointees at the top. As late as the mid-1970s, two thirds of senior positions in Washington were filled by civil servants; now only a third are. Would you want to be an American diplomat if you could never become ambassador to China or the Court of St. James's?

Given the poor pay and hostile atmosphere, civil servants have to look for their status elsewhere. "Parkinson's Law" survives across the West. Visit any cabinet department in Washington, DC, and you will find a bunch of flunkies with inflated titles like chief of staff, "deputy deputy assistant secretary," and, presumably, deputy to the deputy deputy assistant secretary. The Italian president has nine hundred underlings, eight times as many as the German president. There are thousands of state limousines. In Brussels, bureaucrats you have never heard of have bodyguards, cars, and luxurious residences.

7. THE LEADERLESS STATE

Blaming it all on the civil service is wrong. The bigger problem has to do with political leadership. Put simply, not enough good people go near politics. The change from the 1950s and 1960s is especially noticeable. As we saw in the last chapter, the old establishment was convinced that good government mattered: mess up government and you end up with the Depression and Hitler. Now the elite is much less inclined to go into public service.

Money may now be made in much less brutal ways than it was in the days of steel, coal, and slavery. But back then, there was a compensating sense of duty. Go to the churchyard of any British village and you will find a roll call for the dead in the two world wars, normally headed by the squire's son. Even a couple of generations ago, most young men in Europe did a version of what was called national service: they had to serve in the armed forces for a couple of years. Good, many will say: less militarism. But it has removed a link between the wealthy and both the state and the working class. One former British spy points out that his children are immensely better educated than he was, far more tolerant, but the only time they meet the working class is when their internet order arrives; they haven't shared trenches with them. They have been relieved of guilt.

The closer you get to the summit of the global elite the truer this is. People in Silicon Valley do "give back" but they focus on philanthropy outside the public sector. In a knowledge economy, the brain of a Bezos or a Buffett is their most valuable asset. The only obvious member of the West's super-rich to have jumped into full-time public service—Mike Bloomberg—employs one of us; even his foes would say he ran New York City well. Very few have followed him.

The collapse of the party system has not helped. For much of the twentieth century, parties provided a ladder of promotion for able people who wanted to get into politics. The parties of the left in particular groomed a succession of working-class leaders, such as Ernest Bevin, who became Britain's greatest postwar foreign secretary despite leaving school at eleven to work as a draper's boy. Now long-established parties such as France's Socialists and Greece's Pasok are more or less disappearing. Boris Johnson leads a party of 150,000 Tories; his hero Winston Churchill could call on three million. More people vote in celebrity talent shows than in many elections.

This helps explain populism, but it also explains why, in normal times, politics is the preserve of a narrow class who regard politics more as a profession than a calling: politicians who have had their eye on a seat since university, if not before, and a host of other careerists (pollsters, spin doctors, election agents, speech writers, psephologists, and the rest). This new elite is hardly any closer to the real world than Bobbety Salisbury was; the marquis after all had been to war. Its members come from a narrow range of universities and, if they have worked in the private sector at all, a narrow range of professions. The 116th Congress that was confronted by Covid included 214 lawyers but only 11 scientists among its 535 members.[20] Managerial experience tends to be minimal, often comically so. When Alan Milburn took over as Britain's secretary of state for health, and therefore as boss of the largest employer in Europe, his previous managerial experience consisted of being the part-proprietor of a left-wing bookshop in Newcastle called Days of Hope and known locally as Haze of Dope.[21]

If they share some of the aloofness of the old ruling class, the new ruling class have given up some of the previous virtues—notably restraint. They all rush to jobs in the private sector afterward. The Victorian journalist Walter Bagehot argued that, in order to survive, a political regime needed to gain authority from the citizenry, and then use that authority to get the work of government done. In many parts of the West, that chain has broken down. Politicians are among the most despised professions on the planet along with journalists and real estate agents. A study of what words people associate with politicians discovered that the most common were sharply negative: contemptible, disgraceful, parasitical, sleazy, traitorous. Countries where people still trust politicians, such as Canada and the Nordics, are a shrinking minority.

THE RESULT: DEMOCRATIC DISTEMPER

Put these seven faults together—the overload, the fog, the shortage of talent—and the result is democratic distemper. The more voters want from their governments the more they are disappointed. They turn on public officials for not providing them with services even as they deny those same officials the money they need to pay for those services. They subject officials to contradictory demands: they want them to live in capital cities on paltry salaries but then turn on them if they pad their expenses; they criticize them for wearing suits and ties but then turn on them if they wear sneakers and sweatshirts. They want to get rid of tax breaks, but not their own. They want better people to go into politics, but refuse to do so themselves. The universal benefactor also doubles as a universal whipping boy.

The benefactor-cum-whipping boy now has to deal with the biggest crisis since the Second World War: a virus that has infected millions and wrecked the economy—and in the process accentuated all the faults and failures that we have listed above.

CHAPTER FOUR

THE COVID TEST

Codogno, a town of sixteen thousand souls on the plain of the River Po some forty miles southwest of Milan, will never be a tourist destination to compare with Siena, the home of Lorenzetti's Allegory of Good and Bad Government. It nevertheless offers many of the things that make life in small-town Lombardy such a treat: a piazza, a cathedral, dozens of restaurants that routinely produce food of a quality that eludes grander establishments in the rest of the world. The town's patron saint is Saint Blaise, a fourth-century physician.

On February 18, 2020, a thirty-eight-year-old man with no history of illness walked into the emergency room of a local hospital complaining of fever, a cough, and shortness of breath. He was given antibiotics and went home. But a few hours later he returned, complaining that his symptoms had got a lot worse—his fever was raging and he could hardly breathe. He was admitted to intensive care and put on a ventilator. Italy's "patient zero" hadn't even been to China.

Three days later, on February 21, the local authorities reported sixteen cases in Codogno; the next day it was sixty, and five elderly people had died. By February 23, Codogno and other infected towns were turned into "red zones": citizens were told to stay at home, outsiders were prevented from coming into the area. Bars, restaurants, shops, and even churches were closed. Public gatherings were eventually banned across Lombardy. Milan's two secular temples—the La Scala opera and the San Siro football ground—both fell silent. Trains refused to stop at Codogno. But people in a free society are hard to control: thousands of Lombardians fled south in order to escape from the virus, and many of those who stayed in the north evaded the lockdown. The number of infections continued to mount—and along with it the death toll.

The Italian government imposed curbs on large gatherings in the north on March 8, added nationwide limits on travel on March 10, and then closed down restaurants and bars on March 12. The entire Lombardy region, with a population of sixteen million people, was placed in quarantine, with people fined if they left their homes without a certificate explaining why they were doing so. By then, however, the virus was out of control and Lombardy had entered a period in hell. Hospital beds ran out, as did ventilators and protective equipment. The descendants of Saint Blaise, many of them on the edge of exhaustion, had to make life-and-death decisions about who to save, leaving others to die—and die slowly, painfully, and alone, given that relatives couldn't touch the infected without becoming infected themselves. Only the sound of Italians singing to each other from their balconies brought comfort.

Four months later, around 35,000 Italians had died, half of them from Lombardy. In Britain, by midyear the figure was 44,000, while the United States had passed 125,000. And the real numbers could be higher. A study by the *Financial Times* put the number of "excess deaths" in Britain compared to a normal year at closer to 60,000.[1] At the end of June, with the virus surging in Arizona, California, Florida, and Texas, America's government scientists admitted the infection rate could be ten times higher than the official number.

Across the West there were scenes of unimaginable bravery and a surge in communitarian feeling: millions of people went out onto the streets weekly to applaud health workers (with even the Rolling Stones appearing remotely to serenade them). But the suffering, both physical and economic, was staggering. By midyear one in ten Italians and one in five Americans were seeking employment, and Britain was heading toward its worst recession in three hundred years.[2] Even countries that had escaped the virus relatively unscathed were hurt economically: Australia was in its first recession in thirty years. The best guess from *Bloomberg Economics* is that the virus will end up taking $6 trillion out of the global economy.[3]

BLAME BEIJING AND IGNORE THE NUMBERS

There are two ways to explain this pain. The first is to blame the dictatorship where it began. Some claim that the virus came from a government laboratory in Wuhan and somehow escaped into Huanan market, a live-animal and seafood wholesale market. "More and more we're hearing the story," Donald Trump said in April. "We'll see." A fairer charge against China is that a more open country would have contained the virus much earlier. In late December 2019, an ophthalmologist in Wuhan named Li Wenliang privately alerted medical friends via WeChat that a terrible new virus was on the loose. When his online posts leaked, the local police arrested him for rumormongering and forced him to sign a letter of apology. Wenliang later died from the virus, making him Covid's first political martyr. Wuhan's functionaries were frightened—mere cogs in a government machine that thrives on blame-shifting. Even when Beijing finally found out about the virus, China did not rush to alert other countries, letting international flights from Wuhan continue after internal ones stopped. According to estimates from the *South China Morning Post*, from December 30, 2019, to January 22, 2020, eleven thousand people flew from Wuhan to Thailand, almost eleven thousand to Singapore, over nine thousand to Japan, and seven thousand to Hong Kong.[4]

Be this as it may, China can't bear all the blame. It was far less secretive than it was earlier this century with SARS, where six months passed between scientists spotting in November 2002 that a virus had jumped from animals to humans and April 2003 when the obfuscating minister of health was sacked. This time China identified Covid's gene sequence quicker and spread news about it. On January 24, for example, a Chinese medical team published an article in the *Lancet*, a British medical journal, with a detailed description of the symptoms of Covid-19 and warned that a third of patients had to be admitted to intensive care.[5]

The second place to look to explain the crisis, then, is the West itself. The arrival of the virus was a test. Angela Merkel and Jacinda Ardern passed triumphantly in Germany and New Zealand, but most of the West struggled, especially compared with Asia—and Donald Trump and Boris Johnson flunked the exam. Despite the horrors on display in Codogno, too many Western leaders wasted March in dithering or hurtling off in the wrong direction. Then they failed to find the right materials to protect their doctors and nurses and test their populations. Even as late as June, messages remained mixed, with the president of America, maskless, summoning his faithful to a political rally while his scientists warned of cases accelerating to a new record.

Western politicians can question some of the Asian numbers, but even Germany, the best-performing big country in Europe, had a mortality rate that was roughly twenty times as high as South Korea and Japan. On paper China was a hundred times safer than the United States and two hundred times safer than Britain; even if China is as usual cheating, the gap is too immense to disregard.

Exams for governments are not unlike those of students: those who have put in the work and taken the subject seriously tend to do best. To anybody who has studied government competence, there were a few surprises (Greece, under a new reforming government, outperformed; Singapore by its standards did worse because it failed to track migrant workers) but most of the results were entirely predictable. The only striking thing this time round is that people really noticed. Opinion polls showed people in places like Vietnam, Taiwan, and Australia applauding their rulers, while Britons, Americans, and the French have been left dejected.

The failures in the West came in many shapes and sizes: the United States, the worst performer in terms of the number of deaths, and Belgium, the worst in terms of deaths per head, are very different countries with very different leaders. And there are still some mysteries in the numbers: why exactly did Japan do quite so well, and Quebec quite so badly? But three things stand out among the failures: a complete lack of urgency; an inability to organize testing and protective equipment; and dysfunctional politics. That is not to dismiss the often heroic work of many public sector workers, but if there was ever a case of lions led by donkeys this was it.

WEEKS LOST, LIVES LOST

The lack of urgency was criminal. One of the many ways in which Sir Keir Starmer, the Labor Party's new leader, has distinguished himself from Jeremy Corbyn has been to ask Boris Johnson different versions of the same blunderbuss question: why was he so slow? The same question could be put to many leaders across the West.

By the end of February, two things were clear to any Western government that wasn't asleep at the wheel. The first was that the virus had left China and was "going global" (a phrase that *The Economist* put on its cover on February 27). With cases reported in more than fifty countries, the S&P 500 fell sharply. The horrors that were occurring in Codogno were plainly headed toward Cardiff and Chicago. The second was that there were good examples of how to deal with the virus. Since at least the Black Death in 1347, governments have used quarantines to halt the spread of plagues (the term comes from the Italian *quaranta* meaning forty days) and sensible ones did so again with Covid-19. China imposed its strict lockdown on Wuhan on January 23, cutting off all transportation links, and then extending mass quarantine to thirty-six million people across thirteen additional cities soon afterward. The number of new cases tumbled from 3,887 on February 4 to 139 on March 4.[6] Singapore started taking the temperature of air passengers arriving from China on January 22. South Korea advised inhabitants of Daegu and Gyeongbuk, two early hotspots, to stay at home in the middle of February—and closed schools, airports, and military bases.

The West ignored this. Populists predictably rushed to blame the virus on immigrants. Hungary started turning back asylum seekers. Italy's Matteo Salvini of the Northern League linked the disease to the docking of the *Ocean Viking*, a humanitarian ship with a couple of hundred African migrants aboard, none of whom had obviously come from Wuhan. He demanded an "armor-plated" border and challenged Prime Minister Giuseppe Conte to "defend Italy and Italians." Europe's centrists were more grown-up, but dragged their feet. Emmanuel Macron advised the people of France to display "individual and collective discipline" on March 12, but then allowed municipal elections to go ahead on March 15. The French state finally stepped in to impose a lockdown on March 17. That was still a week ahead of Britain, where Johnson continued to go about business as usual, shaking hands with everybody (which eventually landed him with a life-threatening dose of the virus) and failing to attend the first five emergency cabinet meetings on Covid. He pursued a "herd immunity" strategy—keeping schools open, letting mass events like the Cheltenham Festival go ahead, and resisting calls to close airports. More than eighteen million people flew into Britain in the first three months of the year, including hundreds of flights from Wuhan.[7] Johnson only ordered a lockdown on March 23 when a study from Imperial College, London, warned that herd immunity might cost half a million lives as it overwhelmed the NHS. If the lockdown had been imposed a week earlier, one study reckons that three quarters of the Coronavirus-related deaths might have been avoided, putting Britain's death toll on a par with Germany's.[8]

Sweden controversially kept things open even longer than Britain: Swedes continued to frequent bars and restaurants and send their children to school even while their neighboring Nordics had shut everything down. This proved costly in terms of lives, with a death toll of 430 per million, far higher than its neighbors, without saving its economy much pain. Anders Tegnell, the epidemiologist behind this policy, later admitted that he should have pursued a more mainstream strategy.[9]

At least Sweden and Britain followed the experts, if not always the right ones, and admitted their mistakes. Donald Trump did neither. Thus Covid was "totally under control" (January 22), pretty much "shut-down" (February 2), destined to "disappear" (February 27), "the new hoax" (February 28), and "going to go away" (March 12). As Easter approached he advised Americans to fill the shops and churches, while indulging in conspiracy theories about the Deep State inventing the crisis to "bring down Trump" and boasting about his TV ratings ("'Monday Night Football' type numbers"). He ridiculed technocratic experts at both the World Health Organization and his own Centers for Disease Control and Prevention (CDC), trusting instead his own "hunches" that led him to more adventurous solutions, such as controlling the virus with ultraviolet light or injecting detergent. His one attempt to limit his country's exposure predictably involved keeping foreigners out: he stopped flights from China in February and then suddenly flights from EU countries on March 11 (though he excluded Britain, Europe's biggest airport hub). An analysis by Columbia University suggests that, if the United States had started social isolation on March 1, then 83 percent of the deaths in the first wave would have been prevented.[10] As summer started and cases increased again in the South and the West, Trump still refused to wear a mask or desist from political rallies.

To be fair, Trump was not the only American politician who sent unhelpful messages, or engaged in political sniping. In Democratic New York, Mayor Bill de Blasio greeted the virus's arrival by advising people to go and eat in Chinatown. The health commissioner, Oxiris Barbot, told New Yorkers on March 2 "to go about their daily lives, ride the subway, take the bus, go see your neighbors."[11] Having closed schools on March 15, de Blasio talked about a lockdown; the governor, Andrew Cuomo, retorted that it was his decision—and he had "no interest whatsoever or plan whatsoever to contain New York City." By the time the stay-at-home order came on March 20, there were four thousand new cases every day in the city.[12] Cuomo gradually performed better, but as in Britain, sick elderly patients in New York were sent back to care homes. As Fareed Zakaria has pointed out, the core of New York's failure when set aside similar-size Asian cities, with crowded subways and the rest of it, was simply bad government.[13]

Despite having months of warning that the virus was coming, the United States did not really start to treat the pandemic seriously until April, and even then the health guidelines were confused, with, for instance, the CDC going backward and forward over whether the virus could be transmitted through touching contaminated surfaces. But by then in cities like New York the virus, not the government, was in control.

TEST AND YOU SHALL FIND

Moving to lockdown quickly was not an immediate cure-all. One of the more successful countries, Germany, was relatively late, while Japan emerged from only two months of lockdown with fewer than a thousand deaths. The best sources of resilience were testing and tracing, and having enough protective gear.

Asian countries demonstrated this early on. The more clearly you can identify who has the disease, the less you have to depend upon indiscriminate lockdowns. South Korea's government obtained a full-membership list of the secretive church where the virus first appeared, ordered the worshipers to self-isolate, and identified anybody who had been in contact with the infected, using a combination of cell phone surveillance and personal interviews.[14] In India, one reason why Kerala has outperformed the nation is because a dynamic health minister, drawing on her experience with an outbreak of Nipah virus in 2018, started a program of testing, tracing, and isolating at the end of January: by midyear a population of thirty-five million extremely poor people had recorded under ten deaths.[15]

Technically, testing is less important than tracing. Japan succeeded in halting the virus's first attacks without mass testing partly because it had a system of local public health centers, staffed by fifty thousand public health nurses who were particularly good at tracing infections such as influenza and tuberculosis and did the same with the virus. But mass testing obviously makes tracing easier—especially tests for antibodies because they pick up those who have been infected and recovered as well as those who are currently suffering from the disease. Gabriel Leung, an academic at the University of Hong Kong and a member of the WHO team, summed up the message nicely early on: "Test and you shall find. You either test and find it early, and do something about it, or the body bags are going to pile up."[16]

The shining example of this in the rich world was Germany, which started testing and tracing (and then isolating infected patients) in February. Unlike Britain, it deployed private laboratories and, from the first, allowed lay people to administer the tests, which are fairly simple. It also managed the disease locally. Switzerland and Austria were also to the fore, while Finland had big medical stockpiles. Sadly, many other European countries did not have enough tests and the chemical reagents needed. Again, Britain introduced a system of "test, track, and trace" earlier than most European countries, but then had to abandon it on March 12 because it didn't have enough testing kits. Ever the monopolist, Public Health England insisted on supplying all the needed tests itself, rather than tapping the resources of businesses and universities.

The saga of testing in America was especially shambolic. By March 1, at a time when South Korea was testing 10,000 people a day, America had tested 472 people. The CDC rejected the WHO's testing kits and insisted on using its own tests—which turned out to be faulty. It also fought with the Food and Drug Administration and, of course, Trump, who eventually brought in two rival internal task forces to address the testing problem—one run by his son-in-law, Jared Kushner, and one run by the vice president, Mike Pence. Things improved, but by midyear, many poor Americans still found tests hard to come by; by comparison, China had tested all eleven million people in Wuhan, whether they had symptoms or not.

The amateurism with tests was repeated with other forms of medical equipment. In early March, Germany, Russia, Taiwan, and Thailand restricted exports of masks, gowns, and ventilators to protect their domestic supplies, leaving the unprepared to rely on Turkey and China, whose goods often turned out to be shoddy. In New York State, government ventilators were often so old that patients' lungs collapsed; others didn't work at all.[17] In Britain as well, nobody had factored resilience into the public health system. For all its heroics, the National Health Service, the world's largest medical employer, which had done scenario planning for pandemics, didn't have enough masks or gowns. Thousands of volunteers got out their sewing machines and tried to sew gowns and masks. The equipment situation only began to improve when Lord Deighton, the businessman who rescued the London Olympics, took over.[18]

This amateurism reached all the way into Downing Street. The civil service had no contingency plans for what happened if the prime minister fell ill. Not that there was a great selection. Johnson's cabinet was a clique of partisans defined by their willingness to support Brexit "do or die" rather than any particular competence. Johnson refused to bring back Jeremy Hunt, the runner-up for his job who had run the health service for six years. (Hunt argued for testing and tracing from the beginning and constantly urged the government to learn from Asia.) By contrast, when Dominic Cummings, Johnson's closest aide, broke the lockdown laws twice—by driving 270 miles from London to Durham and then driving 30 miles from his parents' house to Barnard Castle "to test his eyesight"—he was forgiven. This stubborn loyalty did as much as anything to destroy trust in Johnson's government and national solidarity. Newspapers offered Cummings masks to their readers on the basis that if you wore one, you could do anything you liked.

EVERY MAN FOR HIMSELF

Indeed, for all the Churchillian rhetoric about fighting the disease together, perhaps the saddest thing about the West was the complete lack of unity. A dozen years ago, as the financial crisis began, the key finance ministers met in Washington, DC, and ushered in a global plan of simultaneous rate cuts, with both the United States and China pumping money into the system. Europe and America cooperated particularly closely, alongside their central banks and multinational institutions, especially the International Monetary Fund. Gordon Brown and Angela Merkel were in regular contact with first George W. Bush and then Barack Obama. Technocrats were to the fore—notably the American trio of Ben Bernanke, Hank Paulson, and Tim Geithner, who guided Bush through the crisis, and Jean-Claude Trichet and then Mario Draghi at the European Central Bank.

This time, there was no "committee to save the world." The first substantial economic response—the Federal Reserve's decision to cut interest rates on March 3—seemed unilateral. There were few signs of cooperation with either other central banks or the White House: indeed Trump reiterated his regret at appointing Jay Powell to run the Fed. The White House blamed the "foreign virus" on China. "This is Covid-19 not Covid-1," Kellyanne Conway, one of Trump's advisers, pointed out, knowingly. When the president banned flights from Europe he didn't bother to give his allies advance warning. He actively considered sending troops to the Canadian border, the longest nonmilitarized border in the world, before bowing to Canadian objections.[19] In June, he cut off American funding from the World Health Organization, an astonishing thing to do in the middle of a global health emergency.

Far from leading the West, the United States disunited in front of the virus. Trump routinely accused the Democrats of inflaming "the Coronavirus situation" for partisan gain and used federal help as a way of punishing them. "If they don't treat you right," he told the White House press corps on March 27, referring to state governors, "I don't call." Democrats responded by accusing Trump of letting blue-state voters die. The virus rapidly became a partisan issue, with Republicans much less likely than Democrats to trust scientists, to consider the virus dangerous, or to do much to avoid contracting it (a remarkable fact given that Republicans are, on average, older than Democrats). Jerry Falwell Jr. informed Fox News viewers that Covid-19 was designed by North Korea to hurt Trump, citing the authority of a waiter who had passed on the news to him. Right-wing militias protested against the lockdown and even invaded the Michigan state house carrying submachine guns.

The European Union failed in its usual way—by dithering and then briefly bursting into life and dithering again. The new European president, Ursula von der Leyen, hardly mentioned the virus in her March 9 press conference reviewing her first hundred days in office. For several months, the countries of the European Union preferred to compete rather than collaborate, with Germany in particular grumbling about resources going to the South. At the central bank, Christine Lagarde found herself stuck in a game of chicken with her governments, publicly urging them to add fiscal rules even as they pushed her to loosen policy further. The initial responses came almost entirely at the national level: countries forgot all the high-flown rhetoric about an "ever closer Union" and focused on their own citizens, abolishing freedom of movement, closing their borders, and imposing lockdowns according to their national timetables.

The north-south divide in Europe widened, worsened by the fact that the South was hit hardest both in economic and also medical terms. António Costa, Portugal's prime minister, declared that "either the EU does what needs to be done or it will end." Italy felt especially deserted. A poll conducted in early March found that 88 percent of Italians felt that Europe was failing to support

Italy.[20] Giuseppe Conte, the Italian prime minister, warned that if the EU didn't "support the entire European economy" during the pandemic, then "the whole EU project would lose its raison d'être." The atmosphere turned so fratricidal that Jacques Delors, the former European Commission president, broke his usual silence to warn that it posed a "mortal danger to the European Union." In April, von der Leyen apologized to Italy for the Union's failure to do more to help it.

Meanwhile, the other divide, between East and West, became more poisonous. Eastern Europeans, fed up with lectures from Brussels on democracy and corruption, turned to Beijing. The Czech president, Miloš Zeman, proclaimed that "China was the only country that helped us." Viktor Orban remarked in a radio interview that he wasn't looking to the EU for help, "because that doesn't work," saying, instead, that he had "people in airports from Beijing to Shanghai." Comparing China's willingness to donate medical equipment with Europe's refusal, the president of Serbia, a prospective EU member, denounced European unity as a "fairy tale."

THE UNTRUSTFUL LEGACY

As spring turned to summer, countries came out of lockdown, and the instructions became more complex, trust became ever more important. Everybody can understand "stay at home or you die"; it is harder to persuade people to stay two meters apart and stagger their use of the subway if they don't believe you. Throughout the Coronavirus test, governments that were trusted got cooperation with testing and voluntary compliance with lockdowns.

Sometimes the planning and thinking was a little too precise: when Switzerland rather wonderfully announced that it would allow prostitutes to go back to work in June but still ban judo, wrestling, and contact sports, it was rightly ridiculed.[21] But in a way the last laugh was really with the Swiss. Here was a government that took the virus seriously from the beginning, collected equipment, tested, tried to give precise instructions to its citizens, and took a realistic view of human nature. The economic plan to protect the Swiss economy was formed by the government and the banks. There were mistakes—it left the border with Italy open longer than the one with Austria—but there is a reason why Switzerland had fewer than two thousand deaths.

A rule emerged. In countries where people trusted their rulers, they took their advice, being willing, as nine out of ten Dutch people told a poll in March, "to give up some of their individual freedoms to keep the Coronavirus from spreading." By contrast those states without that build-up of goodwill had to rely on coercion or fibs that only lessened trust (such as governments claiming that masks did not work—largely because they did not have enough masks). Only about a third of Americans trusted Donald Trump's medical advice.[22] One poll in late April showed that 62 percent of the French had no confidence in their government's handling of the crisis, with commentators, on both the right and the left, comparing France's response to Covid to the country's "strange defeat" by Germany in 1940.[23]

At its worst, this distrust created conspiracy theories: that the virus had been deliberately manufactured, either by China or Big Pharma or indeed the United States; that it spread through 5G towers and masks; that it was a plot to kill off the old. Bill Gates was blamed, because long before Covid he had (correctly) warned about the danger of a global pandemic in a TED talk, and invested cash in trying to find a cure. This nonsense has consequences: people have burned down scores of 5G towers, including sometimes towers that served medical facilities. A third of Americans say that they won't get themselves vaccinated if one is found.

OVERLOADED—AND OVER?

Meanwhile, in terms of geopolitics, the crisis has left the West weaker and Asia stronger. The Western media is full of articles and broadcasts marveling at the way that South Korea, Taiwan, and Vietnam have outperformed their former masters. What made the region so successful? Is it its Confucian tradition? Or its experience of SARS? Or its technological successes? Or is it just much better at running a modern state? There is less global admiration for China, given its role in the virus's origins and the clampdown in Hong Kong. Yet all but the most Trumpian observers concluded the Middle Kingdom did a much better job in terms of protecting its citizens than the other superpower, the United States. Given where the two countries stood a quarter of a century ago, that is no small achievement.

By contrast, the West looks much weaker than it did even a year ago. Look at all the failures we listed in the previous chapter, and the Coronavirus exposed every one of them. Even if a few European countries performed well, the European Union turned in on itself—and will probably be stuck in debates that will be incomprehensible to outsiders for years to come. Shrinking (thanks to Britain's departure), divided (thanks to profound economic strains), and compromised (thanks to Orban and company), the European Union can no longer claim, as it did in the 1990s, to be advancing liberal values.

The United States has fallen further. Ever since the Second World War, America has been the engine of Western success. It organized global institutions such as the World Bank and the IMF. It led the airlift that saved Berlin. It stood up against the spread of Communism in Europe and Southeast Asia. For all its failures in Iraq, it nevertheless led an international coalition against Saddam. With the Coronavirus, it has looked weak, ineffective, and, frankly, weird. What should the rest of the world make of a country whose leader suggests injecting bleach to counteract the Coronavirus? Covid was a devastating epiphany for the United States—a moment when a country "long accustomed to thinking of itself as the best, most efficient, and most technically advanced society in the world is about to be proved an unclothed emperor," as Anne Applebaum has put it.[24]

"Empire" seems an appropriate word. Edward Gibbon's verdict on the Western imperium as most of it emerged from lockdown in July 2020 would be clear. The West had palpably failed the Covid test, in much the same way that Ancient Rome also had dramatic reversals, such as the reigns of Caligula and Nero. Worse, the West's failure in 2020 was entirely predictable—its organs of government had atrophied. Ancient Rome too had these alarm calls when its decline was manifest for all to see. Gibbon would, however, stress one thing: the failure itself is less important than the reaction to it. Rome never woke up and restructured itself. What will the West do?

CHAPTER FIVE

THE MORBID SYMPTOMS

Writing in his *Prison Notebooks* during the Great Depression, the Italian Marxist Antonio Gramsci defined our times: "The Crisis consists precisely of the fact that the old is dying and the new cannot be born. In this interregnum a great variety of morbid symptoms appear." The Western state has been stuck in an interregnum since at least the turn of the century if not the 1970s, but the tendency to produce "morbid symptoms" has been accelerated by the Coronavirus. In Chapter Six and in our conclusion, we will return to the question of what we think should be done to improve the state. But there are all sorts of other ideas being thrown about at the moment—and there is a very high chance that a few of the worst morbid symptoms might be embraced.

Some people see only a narrow political problem: the useless-ness of a particular sort of democratic leader (nearly always, but not only, Donald Trump). Others sense a deeper change. Does it mark a moment when autocracies (nearly always, but not only, China) finally proved their superiority to democracies? For those who would rather stick with democracy, the most popular conclu-sion is that more government is needed. Many sense a move to the left, accompanied by a strong dose of nationalist self-sufficiency. None of these augur well.

NOT JUST TRUMP

The laziest answer is simply to blame the underperformance on a particularly lousy set of leaders. Many establishment-types have rather enjoyed the sight of Trump and Johnson getting their come-uppance and concluded that the people will make wiser choices in the future. As one central banker privately puts it, the populists are finally being forced to accept "the basic tenets of economics, science, and modern diplomacy."

It is certainly true that the Covid crisis was almost laboratory-manufactured to demonstrate Trump et al.'s failures as leaders. Most populist leaders rely on instinct rather than planning, and bluster rather than project management, none more so than Trump. As the virus struck, Trump was on his fourth chief of staff, his fourth national security adviser, and his fifth secretary of the Department of Homeland Security, with both the DHS secretary and his top deputy serving in an "acting" capacity; there were also seventy unfilled positions at the DHS. Other populists—especially Brazil's Jair Bolsonaro—also made a hash of it.

Hence the theory that beguiles many Americans: that Trump is one-of-a-kind. Put a half-capable politician in the White House—for the sake of argument one named Joseph Robinette Biden—and America will do just fine. Beguiling but wrong. First, many of Trump's failings are simply normal politics writ large. The Obama White House was strewn with unfilled desks, while George W. Bush put Michael Brown, a former commissioner for the Arab Horse Association, in charge of the Federal Emergency Management Agency, and even congratulated him ("Heckuva job, Brownie") as he mangled the government's response to Hurricane Katrina. Second, regardless of who was in the Oval Office, the Covid crisis would have exposed the drawbacks of America's health-care and welfare systems.

America's fee-for-service medical system is superb at providing the rich with what they want, whether it be plastic surgery for sagging chins or heart treatment for elderly tickers. It also produces, in aggregate, the world's best medical research. But when it comes to general public health it is a giant sieve. Some twenty-eight million people still lack any health coverage and many more have to spend thousands of dollars on co-payments. The life expectancy and infant mortality figures are appalling for a rich country. Covid struck a system which had little spare capacity, particularly when it came to beds for respiratory diseases, and almost no coordination. And the welfare system made things worse. Even with normal flu, guaranteed sick leave can cut the spread by 40 percent.[1] But around one in four Americans don't normally have any access whatsoever to paid sick leave. So when Covid struck, they stayed at work, infecting their peers. Trump did not create any of this—and to be fair to him even signed a temporary law making employers offer two weeks' paid sick leave if people had childcare problems.

The same goes for any attempt to blame Europe's problems solely on its current leaders. You can argue that Emmanuel Macron was late to spot the virus, that Italy's leadership was especially dysfunctional, or that Covid caught the EU at a particularly weak point, having just changed its commission. But Italians have long complained about the Germans' indifference to their economic suffering: their northern neighbors' failure to provide protective equipment was merely an escalation of that. A different set of leaders might have done a little better—especially in America and Britain. But the West has been too bad at government for too long to blame it all on them.

IN THE NAME OF GOD, GO— AND LET ME TAKE OVER

In April 1653, two years after Hobbes had published *Leviathan* and returned to London from his French exile, a frustrated Oliver Cromwell stormed into the Rump Parliament. Cromwell had won the civil war in parliament's name and executed the king. But now the parliamentarians were doing nothing but plotting and bloviating. The great general had had enough. "You have sat too long for any good you have been doing lately," he supposedly told his fellow MPs. "Depart, I say; and let us have done with you. In the name of God, go!" It was time for a man who could get things done. By the end of the year, Cromwell was formally declared Lord Protector—and was governing without any checks and balances.

The most worrying argument to emerge from the Covid crisis is that enlightened autocracies are better at dealing with problems than chaotic democracies. Naturally the main proponents of this are the autocrats themselves and their flunkies. As early as March, *Global Times*, an English-language newspaper in Beijing, was crowing that China had outperformed the West because "western political systems lack such an ability to mobilize and organize on such a massive scale."[2] When protests broke out in America, crowing turned to ridicule, with the Chinese retweeting "I can't breathe," the last words uttered by George Floyd. Hu Xijin, the editor of the *Global Times*, taunted the House speaker, Nancy Pelosi, and secretary of state, Mike Pompeo, for encouraging the protests in Hong Kong. "Should Beijing support protests in the US, like you glorified rioters in Hong Kong?"[3]

The core of the autocracy argument is not just that China was better organized to protect its citizens but also that it was willing to add the mailed fist to the surgical glove, adopting methods that democracies shy away from. Thus in Wuhan, the authorities isolated anyone who had even the slightest contact with Covid—and effectively imprisoned them in schools, hotels, and sports stadi-ums. When the virus returned in June, in a food market in Beijing, the government again quickly adopted "wartime measures." For China's fans in the emerging world, this brutal efficiency under-lines its superior ability to enrich and protect its people. If you were born poor, would you rather stay poor in a chaotic democ-racy like India or get richer in an autocracy? Many voters know how they feel—not least in India, where Narendra Modi has more than a touch of Oliver Cromwell about him, with his new model army of saffron-clad Hindu nationalists. Predictably, Modi and his acolytes used Covid as a pretext to take on some extraordinary powers: several Indian states claimed the right to arrest people without charge.

Indeed, autocratic-minded leaders have taken advantage of the crisis to insist on a more efficient government, led curiously by themselves. Elections in sixty countries have been delayed. On March 30, Hungary's parliament issued a "Coronavirus law," giving Viktor Orban almost unlimited powers to rule by decree. (To the relief of the EU, he gave them up in May, partly it seems because his majority means he can do what he likes anyway.) The rulers of Serbia, Togo, and Cambodia also awarded themselves the power to rule without constraint. A spokesman for Rodrigo Du-terte, the militaristic president of the Philippines, explained why the virus was best dealt with by the army in Cromwellian words: "[Generals] abhor useless debates. They are silent workers, not vo-racious talkers. They act without fanfare. They get things done."[4]

So great has been the apparent success of the autocrats that it has prompted a degree of soul-searching in the great democracies. The *New York Times* asked whether Europe's high death count was "the price of an open society."[5] Writing in the *New Yorker*, Gary Bass, of Princeton University, pointed to the way that the plague destroyed Athenian democracy, claiming the life of Pericles, its most eloquent proponent. "The model of how democracy began," says Bass, "is also a study in how it can founder and fall."[6] Britain was slow into lockdown not just because Boris Johnson was incompetent but also because of his libertarian instincts (Johnson has a bust of Pericles on his desk). In France, Macron went ahead with the elections that helped spread the virus, partly because he thought it was the right thing to do.

But wait. If there was ever a test that autocracies ought to ace, it is a global pandemic. The problem is what these draconian administrations do to you the rest of the time. Moreover, there is actually nothing antidemocratic about giving elected officials more power in the short term, in order to beat a disease—providing those powers are temporary and pragmatic. Governor Cuomo's decision to tell the National Guard to look for ventilators in New York and Johnson's decision to instruct the army to build a hospital in East London were admirable responses to a crisis—not Covid coups. In the Great Plague of London that Hobbes survived, people from infected households were ordered to carry white sticks when they went out for supplies; as the plague disappeared, so did the white stick. The same should happen with surveillance technology.

And have autocracies really been as efficient as they claim? If China had freedom of the press, for example, local officials would probably have been more frightened of the *Daily Mail* than their overlord in Beijing—and thought twice about muzzling doctors like Li Wenliang. Rather than using their emergency powers to fight Covid efficiently, autocrats have repeatedly used them to settle scores with their rivals. Azerbaijan's ruler took to mixing medical and Marxist language, claiming that isolating members of the opposition had become a "historical necessity."[7] Turkey freed many regular criminals to prevent overcrowding, but strangely left political prisoners locked up. In India, Hindu nationalists accused Muslims of spreading the virus deliberately, coining the terms "Corona terrorism" and "Corona Jihad."

More fundamentally, people who embrace the autocracy argument are too inclined to focus just on the fact that America has handled the virus worse than China. That doesn't prove that autocracy is a better answer than democracy. In the battle against the virus, there have been high-performing democracies as well as low-performing democracies, just as there have been high-performing autocracies and low-performing autocracies. Why not compare China with South Korea, Japan, and Taiwan? Or with Germany, Switzerland, and New Zealand? When people think of the West, they should think not just of Donald Trump (who does not care much about the West anyway) but of prudent Mrs. Merkel too.

By contrast, on the authoritarian side of the ledger, Russia has looked weak. Vladimir Putin's popularity has dropped to its lowest level since 2000, as Russia has struggled to deal with both the virus and low energy prices.[8] Venezuela's Nicolás Maduro has the same problem. In Iran the government preferred grandstanding to problem-solving. At one point it sent masks to Wuhan, to demonstrate its solidarity, just a few weeks before it started to dig mass graves at home. If you believe North Korea's claims to have no infections so far, we have a bridge in Brooklyn you might want to buy.

For all the economic advances that China has made, democracies are still much stronger. James Arroyo, the director of the Ditchley Foundation, points out that if you divide the world between autocrats (China, Russia, and so on) and democrats (the United States, the EU, the UK, Japan, and South Korea, for starters), and then you measure their strength in terms of GDP, armed forces, and technology, the democratic world still leads the way. And even if you concede that autocracy can help emerging countries grow for a while, the richer such regimes become, the more their people turn from Hobbes to Mill. Taiwan, South Korea, and even Singapore have become more democratic the more prosperous they have become. Of all the explanations of why Seoul did so much better than New York in handling Covid, the idea that the Korean city has a submissive Asian culture is the most outdated. This is the Asia that globalization has spawned—and that the West needs to embrace.

Our guess is that China's middle classes will want the same liberties. They do not want to return to anarchy; but they want a country in which their voice is heard and the rule of law applies to their leaders. Asked whether China has done well during the crisis, one of America's preeminent sinologists, with close ties to the leadership, agrees that it has been more efficient than his own country; but he also wonders how much longer the Chinese state will be able to enforce draconian measures on their ever richer, ever more liberty-loving subjects.

The virus may help some authoritarian regimes last a little longer provided they deal with it efficiently. It may also allow some populists in democracies to grab some unjustified powers. But it will not convince the people of the West to give up on elections. The bigger question is what will those people now vote for—and we fear the answer is larger, more nationalist governments.

THE EVEN GREATER SOCIETY

The Coronavirus arrived just as Bernie Sanders and Jeremy Corbyn, the two most left-wing people in the Anglosphere's recent history, were bowing off the political stage. Having seen their ideas comprehensively rejected by voters in Britain's general election and the Democratic primaries, and knowing their parties had chosen pragmatic leaders, you might have imagined the old firebrands would be a bit downcast, even humbled. Not a bit of it. Both these wizards of the left exited in giant puffs of self-adulation, claiming that the virus proved that they had "won the argument" about the virtues of high public spending and higher taxation.

For once, most political observers agree with them. One common early conclusion from the virus is that it must entail "more government." Peter Hennessy, a historian and crossbench peer in the House of Lords, argues that history is headed for a more interventionist "filled-in" state: "The tide of ideas and the tide of practicalities have turned. A new consensus is coming out of necessity."[9] There will be a great turn to the left, much as there was a century ago when the world decided to build the welfare state.

If it happens, this will be continuing a trend not bucking one. Before the virus struck, Leviathan had already had two recent growth spurts—the expansion in security powers after September 11, 2001, and in economic powers after the financial crisis: in the decade after 2008, the Federal Reserve and its peers in the European Union, Britain, and Japan injected about $13 trillion into their economies.[10] This time round, it looks like the number will be even bigger: on one day in June the European Central Bank alone committed $1.5 trillion in cheap loans.[11] That excludes the money committed by governments, with even misers like Mrs. Merkel reaching into their pockets to bail out companies. Uncle Sam is already on the line for over $3 trillion in stimulus, deferrals, and guarantees.[12] By the end of May around forty-five million jobs were being supported by governments in the euro area alone. When British shops reopened, a cartoon in the *Daily Telegraph* showed a consumer, loaded down with goods, announcing, "Just charge it to Rishi Sunak," Britain's finance minister.

The left wants more. Corbynistas and Sandersistas are having a merry time attaching all their favorite policies onto Covid-19 like so many baubles on a Christmas tree—radical redistribution here, universal basic income there. "The Corona crisis is not without its advantages," says Ulrike Herrmann, a German anticapitalist. Some of this is silly, but two historically left-wing ideas are gathering admirers across the political spectrum: industrial policy and leveling up.

In the eurozone, even before Covid, there was pressure to create companies big enough to compete with America's and China's behemoths. With airlines being renationalized, ever more workers on the payroll, and all those pesky single-market rules banning state subsidies forgotten, Colbertism is on the increase again. That includes telling companies where they can invest. European Union officials talk of "strategic autonomy." In both Japan and Britain similar moves are being made, especially to protect the economy from the Chinese.[13] A new industrial policy is being forged around the idea of resilience—the need to build up capacity to deal with future shocks. Much better to order companies to store protective equipment and vaccines than rely on a global market that can easily be overwhelmed.

Meanwhile the killing of George Floyd coupled with the far higher death rates for minorities and the poor from Covid has given a boost to the idea that society needs to be "leveled up," especially when it comes to racial disparities. In Minneapolis, the city where Floyd was killed, the median income of Black households is about a third of that of white ones. Nationally, the figure is two thirds. The figures for wealth are even more extreme. In 2016, the last year in power for America's first Black president, the median white family's net worth was ten times the figure for the median Black family.[14] Though other countries in the West are not quite so extreme, the race-effect is still noticeable. In Britain, for instance, the death rate for the virus was 10 to 50 percent higher for most Black and Asian groups, compared with whites; for people of Bangladeshi origin the rate was twice that of whites.[15] The idea that "something must be done" has taken hold—just as it did in both the 1930s and the 1960s.

A decade ago, you might have seen pushback from the right against all these new obligations being foisted on the taxpayer. Not this time. "Even someone like me, the most conservative and fiscally conscious senator in the country, is willing to spend federal dollars to help millions of workers," announced Rand Paul, the libertarian champion in the Senate. For the populist wing of the right, there has been something close to enthusiasm. Johnson's Conservatives won the British election in 2019 on a platform of spending more on just about everything (just less than Jeremy Corbyn). At the end of June, Johnson delivered a speech comparing himself to FDR in his desire to "build, build, build." In America Trumpism rejected decades' worth of Republican orthodoxy on free trade and small government. Steve Bannon, the mastermind of Trump's election victory, has long thought limited-government conservatism is old hat.[16] For him the idea that you can remain dependent on China for vital goods like medical supplies while pushing back against that country's geo-strategic ambitions is absurd. His former colleague in the White House, Peter Navarro, has declared that "never again should we have to depend on the rest of the world for our essential medicines and countermeasures."[17]

Even conservatives are asking: what good was globalization in the crisis? Countries rapidly took an each-man-for-himself approach, shutting their borders and making their own equipment. Where was the European Union in the biggest pandemic for a century? Johnson talked of a return to "national self-sufficiency" in vital medical supplies. After listening to global businesses preaching about the virtues of lean supply chains and so on, politicians have noticed the way the tycoons came running home this year: Richard Branson, a man who lives in tax exile in his own private island, miraculously remembered his loyalty to the British crown when Virgin Atlantic needed handouts.

There is one final reason why even the right accepts that government may grow significantly larger, at least in the short term. Despite fiscal deficits yawning ever wider, most Western governments are able to borrow at close to zero interest rates (or in the case of the most virtuous ones like Germany at negative rates). As Sebastian Mallaby, of the Council of Foreign Relations, has put it, we live in an "age of magical money."[18] Back in the 1990s, James Carville, Bill Clinton's equivalent of Bannon, said he wanted to be reincarnated as the bond market—because it was so much more powerful than the state. Nowadays the bond market is the state's junky, addicted to cheap money that only Leviathan can print.

BEATRICE, YOUR TIME HAS COME

So will there be an expansion of government in the West? Our fear is yes. Given the political winds that support that idea and the tolerance of the markets, it is hard to see anything else, in the short term at least. Nationalist arguments for self-sufficiency and socialist ones for a big state have begun to blend, much as they did in the early twentieth century when Beatrice Webb wanted to build the welfare state and Winston Churchill wanted to contain Germany.

Leviathan's growth has often been driven by pestilence and economic collapse. Plague forced the small-government Victorians to create public health and sewage systems, just as the Great Depression forced laissez-faire America to embrace a system of social security. New programs, once introduced, have a way of putting down deep roots and becoming part of the vegetation, not least because they change people's deepest expectations. Before the Great Depression only eccentrics in America advocated social security. After the Great Depression only eccentrics advocated getting rid of it.

All the same, there are two reasons why an even larger Leviathan will not (and indeed should not) last forever. The first is that, for all their current tolerance, the markets and taxpayers will not stomach it. When governments spent heavily after the financial crisis, a round of belt-tightening quickly followed. Austerity could happen again. Fantasies about universal basic income will not survive an era of postpandemic belt-tightening. Look at France: Bruno Le Maire, the finance minister, has already ruled out tax rises. Or look at Italy, the country that has been most successful in resisting reform in Europe. For the moment, the markets are lending it money. But at some point, the penny will drop (or perhaps it will be told to drop by Mrs. Merkel). Asked "how did you go bankrupt?" a character in Ernest Hemingway's novel *The Sun Also Rises* replies, "Two ways: gradually and then suddenly." Italians may well face the same problem—and when gradually becomes suddenly they will have to make a choice about where their money is spent. Do they continue with the ancestral perks and corruptions that litter their system? Or do they redirect the cash toward their poor and needy?

All of which ties into the second more important reason. There is no need to expand the state overall. Leviathan has kept on expanding for decades without getting noticeably better at doing its job: calls for boosting state capacity have often resulted in state enfeeblement because they have given it too much to do. Why should it be different this time? The size and above all the flabbiness of the state can be a problem rather than a solution. Government can do most of what it needs to do just by redirecting its resources toward the poor and needy and modernizing its methods. Make no mistake about it: in some places, such as health care, we think the state should do more. But Leviathan also needs pruning of self-indulgent government services. What matters is what works.

CHAPTER SIX

WHAT WOULD BILL LINCOLN DO?

One of the West's great strengths has been its talent for reinvention. Just when everything looks hopeless, it succeeds in regenerating itself, spurred on by new ideas, new technology, and the threat of competition. Woken up, the West could do a lot, quite quickly. Our hope is that the pandemic, by exposing so many weaknesses, will force Western governments to embark on a sustained period of reform.

Any renewal must involve three ingredients: basic modernization; luring talented people back into public service; and focusing the state on what it does well. They are linked. An unmodernized state that tries to do everything (and therefore does lots of things badly) will never get good people to work for it—and without better people, the public sector doesn't have a chance of successful reform. In some cases, reform involves big philosophical choices: we will return to those in our conclusion. But most of this, as we have already seen, is simply copying what other organizations—the Western private sector, governments in Asia—have been doing for decades. All it requires is will and determination.

Going through precisely what is needed in each country would require a book even longer than Samuel Finer's 1,700-page three-volume history of government that he left unfinished.[1] So consider the following thought experiment. The challenge now seems most similar to the one faced in the nineteenth century when a new liberal order of open competition and efficiency swept away a flabbier old order of patronage and corruption. Imagine that the two most formidable Anglo-Saxon politicians from that era, Abraham Lincoln and William Gladstone, were resurrected, fused together, and elected to the White House on a platform of reforming government. What could President Bill Lincoln do?

Our new president would combine the best of his two antecedents. Neither were perfect men—especially when young—but once they got to power they showed a willingness to cleanse government. From the "People's William," Bill Lincoln inherits a drive to direct resources away from the Old Corruption of special interests and cash-for-perks toward those who really need them. From "Honest Abe," he gets a desire to unite his country and rid it of the scourge of racial injustice. Both men believed passionately in improving the lot of ordinary men—especially through education. But both

men also believed that Leviathan shouldn't try to do too much—
"Government should do for people only what they cannot do better
for themselves, and no more" in the first Lincoln's words. President
Bill Lincoln could be either a Republican or a Democrat—he is both
a "left-wing" social reformer and a "right-wing" small-government
man who believes in self-reliance. He is a patriot, passionately worried that the United States is falling behind authoritarian China.
And he is deeply committed to liberty around the globe.

So, put the great man to work. Both of Bill Lincoln's antecedents drew from the ancients, especially the Bible, but focused
firmly on the future. So he too is looking ahead. Let him wander
through the dusty corridors of government bureaucracy. But also
let him visit the multicolored campuses of the private sector, and
allow him a whirlwind tour of places that work—Singapore, Denmark, New Zealand, and so on. Give him perhaps a couple of evenings off to go to the theater and to counsel fallen women in the
Willard Hotel (with the press kept at a distance). But above all give
him freedom to work.

These are the thirteen reforms President Bill Lincoln would
start applying to the United States. They are based on what is already working around the world, so the barriers to implementing
them are political, not practical. We chose the United States because it is the West's biggest country, its natural leader, and the
one that has most obviously flunked the Covid test. It is of course
sui generis. But most of these reforms could be applied in other
countries: we will return to them at the end.

1. BUILD RESILIENCE

The obvious place for Bill Lincoln to start is the complete lack of preparedness for the virus. His country has just lost more citizens to Covid than it did in Vietnam. This lack of preparedness was a dereliction of public duty given that the first role of any government is to protect its citizens from threats to life and limb.

It is not as if the United States was not warned. Covid-19 was the third outbreak of a Coronavirus this century—after SARS (2003) and MERS (2012). It had also watched the damage done by swine flu in 2009, Ebola in 2014, and Zika in 2016. For a while America looked prepared. Under Bill Clinton it created a National Pharmaceutical Stockpile to store supplies. George W. Bush warned the American public to beware of a global pandemic, declaring that "if we wait for a pandemic to appear, it will be too late to prepare." Barack Obama drew up ambitious plans to produce twenty million reusable face masks and cheap ventilators. But when the stockpile was tapped to deal with swine flu, Ebola, and Zika, the supplies were never fully replaced. And Trump cared even less. In 2018, the National Security Council's "pandemic preparedness" team was dissolved, and the next year the administration withdrew an epidemiologist it had embedded with China's epidemic unit. Since 2010, the CDC's budget has either fallen or remained flat, depending on who you believe.[2]

The government of the United States was so overloaded with less important preoccupations that it lost the ability to focus on real dangers. The urgent always drove out the important. And once the virus struck, Washington had lost the ability to learn—not only compared with Asian governments, but also compared with America's private sector. The head of one of the country's biggest employers admits that, at the start of the year, he underestimated Covid too, dismissing it as an East Asian phenomenon. But by February, once it was clearly spreading, his team studied how countries had dealt with SARS: they talked to officials, started ordering equipment, worked on how social distancing and stay-at-home requirements would affect them, organized emergency financing, and so on. The CEO still berates himself for not being fast enough, and is trying to "benchmark" his performance against other companies, but he says that he was flabbergasted by the way that government officials in the United States seldom tried to make simple phone calls to find out what worked in Asia.

What would President Bill Lincoln do about this? He would overhaul America's health system to be sure, with the aim of making it cheaper and fairer. More on that later. Rather than the government building up central stockpiles of medical supplies, he could copy the cheaper, more flexible Swiss system, where each employer is responsible for keeping up-to-date protective equipment for its workers.[3] He would certainly rejoin the World Health Organization, not out of some wooly multilateralism, but because a pandemic is a global problem: not working with other countries loses one friends and costs both lives and money. And he would also ask himself: where else is America unprepared to protect its citizens against catastrophe?

Cybersecurity is one threat—but we will leave that to the spies. The other obvious area, where resilience now looks more important than cheeseparing even to this thrifty president, is climate change. Like the pandemic, the exact nature of the threat is debated, but, given the damage a warmer world could do, it is crazy not to take out insurance. The cheapest premiums are multilateral ones: America will get more protection through a global-warming agreement than it will ever get unilaterally. But there are still things for Bill Lincoln to do at home. The $20 billion in subsidies that go to the fossil fuel industry, even though the world is awash with cheap oil, would obviously disappear.[4] But how to encourage renewables? The most efficient way to change behavior is a carbon tax, which hands the job of choosing technologies to the market. Some twenty-five countries already have one, including Japan and Singapore. Lincoln would copy Canada, also a resource-rich country with a powerful energy industry: it introduced a clever carbon tax that started relatively low at $20 a ton in 2019, but rises gradually to $50 a ton in 2022—and stays at that level. As in Canada, individual states could raise it higher—and some of the proceeds could be directed toward workers who lost their jobs, such as coal miners in West Virginia.

2. PROTECT AND UNITE

If the pandemic exposed how an unresponsive Leviathan fails to protect, the murder of George Floyd showed how it can actually do harm. The warnings America had about viruses striking its population were theoretical and sporadic; the warnings about racist policing were real and repetitive. Nearly thirty years after one of us covered Rodney King's beating in Los Angeles, Blacks are three times more likely than whites to be killed by the police; indeed, death-by-cop is the sixth leading cause of death for young Black men.[5]

There are two constitutional problems that bedevil police reform in America. First, most policing is local—far too local. There are more than eighteen thousand law enforcement agencies, some of them tiny, and in big cities like Los Angeles, multiple different forces still overlap. Second, thanks to the Second Amendment, America is a heavily armed country where guns kill forty thousand people a year. The police are terrified of being shot: most of the people they shoot have guns. Lincoln should do anything possible to get guns off the streets and toughen up background checks. More immediately, he could abolish the Pentagon program that doles out its surplus weapons to the police. When police swagger round even small towns with armed Humvees and machine guns that have seen service in Iraq, they look like an occupying army.

The United States needs to fire bad cops. When it comes to racism and violence, armed officers of the law should be held to higher standards, not lower ones, than the rest of us. To get round the unions, federal law should insist that disciplinary records are kept in full. America should follow other countries and require more training for its cops: it's astonishing that a Louisiana cop can use deadly force after just three hundred hours of training. Camden, a rough town in New Jersey, reduced its police violence problem with education in de-escalation and conflict management. More broadly, the job should be redefined. The police are a classic example of the overloaded state, with cops being asked to deal with problems such as mental health, family breakdown, and juvenile delinquency. "Defund the police" should become "deconstruct the police," with some functions handed over to trained (and unarmed) social workers.

Police reform by itself will not right the system that throttled Floyd to death. Gladstone made a habit of creating big commissions into pressing social problems and then implementing their recommendations. Bill Lincoln could set up one to look at criminal justice, particularly America's habit of sending people to prison for minor offenses. The total incarceration rate is roughly double that of Turkey, the sternest in Europe, and eight times the rate in Scandinavia, the Netherlands, and Ireland.[6] There are more Blacks than whites in American prisons, despite the fact they make up only a sixth of the population.[7]

A second reform commission would look at poverty—an obsession for both the original Lincoln and Gladstone. From cradle to grave, Black America gets the worst of the state—shoddy maternity care, lousy preschool education, substandard schools, expensive universities, no sick pay, and a medical system designed for rich people. In some cases, righting these wrongs will involve more public spending, but more cash should be conditional on the producer lobbies—especially teachers unions—accepting reform. The relative success of charter schools and voucher programs could begin to provide Blacks with more control over their own lives. A reformed welfare state that sought to help the poor and needy rather than the old and rich would help Black America disproportionately.

3. LIFT THE FOG

Unless you expose Leviathan to the light, you will never persuade voters to accept change. When Gladstone went into politics, Whitehall's accounts, insofar as they existed, were deliberately incomprehensible. The aristocrats who ran the country wanted to conceal the fact that most government spending went to support their relations in the form of sinecures, church livings, pensions, farm subsidies, and ceremonial jobs. Gladstone insisted on standing before parliament, sometimes for hours on end, and explaining the budget in detail.

As we have seen, America's budget is similarly incomprehensible because it is stuffed with perks and exemptions that benefit special interests. Bills don't run to thousands of pages by accident. They do so in order to conceal what is going on from the public—and indeed from many lawmakers themselves. Washington would benefit from a bit of Gladstonian theater, with the new president promising to explain each line of his budget in the State of the Union, rather than the current soap opera. A true costing would include both an honest reckoning of the guarantees given by government—for instance behind student loans and Social Security—and the money Uncle Sam gives back to the rich through "tax expenditures." At the moment the trillion dollars the federal government spends on health care does not include another $225 billion in deductions for employer-provided health insurance.[8]

If anything, the books are even more cooked at the state level. Take the first Lincoln's home state. Illinois is awash in dodgy numbers, especially when it comes to pensions. In 2019, the Chicago Teachers Fund had a funding ratio of just 41 percent. Because the state has never put in enough money, three quarters of the cash that it gives now just goes toward covering previous shortfalls.[9] The state's official liability of $13 billion depends on it using an absurdly low discount rate. It should use the same formula a company would have to do, which raises the figure closer to $20 billion.

4. SIMPLIFY, CUT, MODERNIZE, SELL

With honest numbers, you can begin to simplify and slim Leviathan. Does America need a tax code that is so complicated that nine out of ten tax filers pay for help to complete their returns? For years, reformers have argued that the income tax rate could be lowered if you got rid of the $1.6 trillion of exemptions which go mainly to the well off.[10] Do it now. Make the tax rates for income and capital gains the same, so richer people are no longer incentivized to shuffle income and wealth from one place to another. Shift more of the burden onto consumption, rather than income: America is the only large country without a value-added tax. And why should poorer Americans fill out a return when all their income is salary? Some forty countries offer "return-free filing," where the tax authorities simply do your taxes for you (in Sweden you can do it on your phone). Around fifty million Americans would qualify for this.[11]

Similarly, remember the perpetually expanding *Federal Register*? Does the United States need a pile of regulations that reaches ever upward to the moon? One way of addressing this problem would be more widespread use of "sunset clauses," so that laws and rules automatically expire. Covid provides a prompt. Several parliaments, including Britain's, have set formal time limits on the emergency powers they granted governments. So far, America's experiments with sunset clauses have been too half-hearted—they have been used as accounting gimmicks to get tax cuts into balanced budgets. But no other organization would commit itself to a set of rules in perpetuity. Sunset clauses are not perfect: you would certainly need exemptions for infrastructure bills or projects that obviously need to extend longer. But why not insist that all new laws and regulations expire after ten years? The only way to address the automatic machine that creates ever-greater complexity is to create another automatic machine that pushes in the opposite direction.

And while you are simplifying things, why not get to work on out-of-date institutions? The Department of Agriculture should finally go: it can be carved up between the Environmental Protection Agency and a food safety agency. And America is not the Soviet Union. National parks and military bases aside, there is no need for the federal government to own so much of the country. Selling some of the 900,000 buildings and 640 million acres of land owned by the government would allow you to cut its size while also producing a revenue windfall.[12]

5. STOP SUBSIDIZING THE RICH AND THE OLD

Getting rid of exemptions in the tax code and simplifying laws is not just a matter of modernization and efficiency. It prompts greater accountability. With the exemptions gone, it will be far harder for special interests to do their dirty business. In a fantasy world, we would love Bill Lincoln to go further and confront the scourge of money politics and American feudalism head on; given the constitutional restraints, reforming campaign finance may have to wait till his second term. But simpler taxes and rules would certainly begin the long process of redirecting the American state's largesse away from the already-well-off to the truly deserving. Every dollar liberated by canceling subsidies for ethanol producers and tax breaks for hedge funds can be used elsewhere.

The elephant in the room is, of course, entitlements. We have already explained how the elderly, not the poor, are the biggest charges on the state. This should be an argument about security. "You and I agree that security is our greatest need," Franklin Roosevelt told Americans in one of his fireside chats in 1938. "Therefore I am determined to do all in my power to help you attain security." That is why he called the pension program he created in 1935 "social security." But the biggest danger to any safety net is to stretch it too thin. America's poor get less than they need because the main benefits are universal ones. It is not clear that the security of Bruce Springsteen or Warren Buffett is in any way improved by the fact they are entitled to a free state pension. Means-testing Social Security and raising the retirement age rapidly to seventy would help balance the budget.

Of course, the politics of this are explosive. But again copy what has worked elsewhere—and set up an independent commission charged with reforming the entitlement system (with its final proposal subject to a straight yes/no vote). Sweden did this in the 1990s with the explicit aim of bringing its entitlement system, which had reached American levels of dysfunction, into balance. All Swedes still get pensions, but they don't automatically increase regardless of the country's ability to afford them.

Too radical for America? This is the country that set up an independent central bank in 1913 to avoid the repetition of the situation in which J. P. Morgan rescued the country from financial collapse by locking bankers in his house until they came up with a plan to prevent the run on the dollar. As Garett Jones points out, giving powers away to independent central bankers has reduced inflation—and saved politicians from having to make unpopular decisions.[13] Doing the same with entitlements could make sense.

6. A FAIRER HEALTH-CARE SYSTEM

Sooner or later, any reforming president has to grapple with the most wasteful part of America's government: its gerrymandered health-care system. Bill Lincoln could argue that the pandemic has made a compelling case for sweeping reform, not just to remove the main source of insecurity for the poor but also to save money. The current absurdly complicated system, with its hidden subsidies, convoluted insurance requirements, and misplaced incentives, means the United States spends around 18 percent of its GDP on health care, far more than anywhere else, while leaving one in five nonelderly people uninsured. And no matter how much the health industry and their lapdogs in the Republican Party howl about the threat of socialized medicine, there is nothing free market about a system that spends proportionately more public money on health than "socialist" Sweden (which, like just about every other rich country, has a healthier population).

There is no perfect health system. One option would be to expand Medicare, the public system for seniors, so it covers younger people too. This is a simple way to extend coverage within the existing arrangements, but, without other reforms, it would add costs. Bill Lincoln could draw on the German model, where health insurance is compulsory, with 90 percent of the population using subsidized public insurance and the richest tenth staying private. Germany manages to keep waiting times relatively short, but with more than a hundred funds to choose from it has some of the complexity that bedevils America. Canada and Singapore both have merits. Drug firms and insurers have done a good job of persuading Americans that single-payer systems, like Canada's, where the state foots the bill, involve long waiting times at government hospitals. Against that, having a single payer with set fees for procedures gets rid of a lot of the paperwork that comes from insurers, patients, and hospitals arguing about who covers what, whether doctors are "in network" or "out of network," how big a co-payment is needed, and the rest of the American health-care nightmare. The administration costs in America's system are roughly double Canada's.[14] The Canadian government also has much greater bargaining power over drug companies, so medicines are cheaper north of the border. The beauty of Singapore's Central Provident Fund is that it makes people more responsible for their health. It also requires users to pay a small fee when they visit their primary provider in order to discourage the unnecessary visits that plague "all you can eat" systems like Britain's NHS.

However Bill Lincoln put his system together, it would combine three features. It would guarantee every American a certain standard of free health care, paid for by the government but provided at both public and private hospitals. There would be a hypothecated health-care tax so each American can see how much the public system costs on their (enormously shorter) income tax return. And he would continue to subsidize private health insurance at the personal level (on the basis that some private spending is saving the state money). Individual Americans would be allowed a small tax break, capped at, say, $500. But this perk would be personal and portable, not the convoluted corporate version (introduced by accident during the Second World War to deal with temporary labor shortages). There would be incentives to stay healthy and get vaccinated. But, as in slimmer countries, there would be taxes on the sugar and junk food that encourage obesity and diabetes. Private medicine would survive, as it does in Singapore, Germany, and Canada.

7. EDUCATE OUR MASTERS

Bill Lincoln would be particularly struck by how far the West has fallen behind in education. The first Lincoln laid the foundations of America's state university system with his land-grant colleges. Gladstone believed in the importance of "educating our masters" for the emerging democratic society. You can imagine Bill Lincoln thumbing through the PISA tests that measure the excellence of schools around the world in reading, math, and science and being horrified by how far America has fallen behind. The same names keep appearing at the top: Singapore, Taiwan, South Korea, Japan from Asia; Finland, Estonia, and Switzerland from Europe, with Poland and Ireland doing well; and Canada.[15]

The United States is only middle-of-the-league despite spending more per head than much more successful countries. (This failure is partially offset by America's success at higher education, but its universities are becoming overburdened by administrative bloat and, in the humanities, intellectual decay.) America's schools should ignore the teachers unions (and their lapdogs in the Democratic Party) and learn from more successful countries. These tend to have several things in common: they pay good teachers well and weed out bad ones; they focus on core subjects; and they provide a variety of schools to address the variety of abilities and aptitudes.

Alongside much stricter standards for teachers and a focus on excellence, there would also be more money in some chosen areas. One is preschool education. This is where inequality starts—where the middle classes who can afford nannies and tutors start to move ahead, and the children of single parents are punished. America's refusal to spend money on it is self-defeating—and shows up in those league tables. The other area of educational investment would be at the other end of life. Technology is changing jobs at a time when Americans live and work longer. Most people end their university education in their twenties. The option of a year's subsidized education once you reach fifty should also be part of our contract with Leviathan. Right-wingers like to argue that overseas aid should be focused on providing the poor with fishing rods rather than fish. The same applies to older Americans.

8. UNLEASH TECHNOLOGY

Again, both the first Lincoln and Gladstone were obsessed by harnessing technology, the former as an inventor (with a patent for lifting boats), the latter as the champion of Victorian entrepreneurs. Back then, railways and telegraphs changed government in both countries, helping them to pull ahead. Today America's lead over China in technology through Silicon Valley is perhaps its most important asset. Yet very little of that inventiveness has been applied to America's public sector. What chance was there of fighting Covid, when around 40 percent of the IT systems at the Department of Health and Human Services are "legacy" ones, no longer supported by their manufacturers?[16]

Asian governments are stealing a march on America in using the internet of things to monitor smart infrastructure. In Singapore, water pipes report back to the authorities if they spring a leak, while lampposts gather data on temperature, humidity, and traffic flow. Some American states are getting better at communicating with people through mobile phones and apps: but again, Covid underlined how far ahead East Asia is. In Shanghai, each subway car has its own QR code (or bar code) that you scan when you get on, so that if one of the passengers gets sick, only people who have traveled in that particular car need to be contacted.[17] Of course there are privacy concerns with this, but the main barrier to this happening in America is technological. The New York subway system only started introducing Asian-style cashless payments in

2019. And on the subject of cashless systems, China is building the infrastructure for a digital currency that some people think might unseat the dollar.[18]

The United States has stinted on high-tech infrastructure for the same two reasons that it has let its bridges and roads crumble: because entitlements absorb so much cash and because nobody counts the dilapidation in the national accounts. America's tech budget is eaten up by the cost of supporting legacy systems—and the elderly workers who run them—because nobody has had the courage to pay for the upgrade. Bill Lincoln should borrow from America's past, as well as Asia's present. Roosevelt built the dams. Eisenhower built the freeways. Bill Lincoln will use the fact that America can borrow long-term money at close to 0 percent to build the infrastructure a knowledge economy needs. That includes a subsidized internet, but also an overhaul of technology in every department. Otherwise the shabbiness of LaGuardia airport will be repeated in cyberspace.

9. GO LOCAL

An important reason why Seoul did so well in dealing with Covid had nothing to do with technology or efficiency. It had a good mayor. Park Won-soon died in mysterious circumstances in early July. But until then he was regarded as a reformer. Elected on an anticorruption ticket in 2011, he had a record of trying to explain problems to citizens and involving them in producing solutions.

His great theme was the "sharing city" initiative—beginning with roads and parks and then moving on to technology. Park even took his family to live in a fetid shack in the poorest part of the city.

As James Anderson, head of government innovation programs at Bloomberg Philanthropies, points out, there is a wave of reforming mayors around the world. Cities are better at government in part because they are closer to their people than national politicians are; they also tend to be much less partisan. When he was mayor of London, Ken Livingstone, a left-wing Labor Party figure, embraced the "entirely capitalist" scheme of road pricing while his Tory successor, Boris Johnson, embraced the "entirely communist scheme" of bike sharing.

One of the more striking things about America is that it has gotten worse at using the federal system to learn. As "laboratories of democracy," the states used to be America's way of renewing itself. Welfare reform and charter schools both began in Minnesota and then spread across the country. Barack Obama established a "Race to the Top" program to spread successful ideas in education; it fizzled out. Bill Lincoln would certainly try to spread ideas around the states, but at the moment they are not where the action is. Cities are. Seattle and San Francisco both did reasonably well at handling Covid (certainly compared with Washington, DC). San Francisco and Boston are good at technology; Dallas leads the way on toll roads; New York has massively improved air quality. After Hurricane Katrina in 2005, New Orleans embraced charter schools and school vouchers.

Since the 1960s the federal government has generally central-
ized power, shifting it to the place that is farthest away from the
people and closest to the organized special interests that crowd
together in K Street. That process must be reversed. Big-city mayors should get more power over schools, transport, and police,
and they should also be encouraged to copy successful ideas from
other cities through a special federal government fund. Their
ideas should be financed, not stymied.

10. REINVIGORATE TALENT

Any overhaul of government has to include improving the quality
of people in the public sector. Prime Minister Gladstone was a pi-
oneer of meritocracy—the idea that you have to put the cleverest
people you can find in charge of the machinery of state. He masterminded the Northcote-Trevelyan reforms that opened jobs in the
British Civil Service to competitive examination. An elite corps
of civil servants succeeded in reshaping the government of both
Britain and the Empire. The French bureaucracy that emerged
from the Revolution and Napoleonic Wars was even more selec-
tive than Gladstone's version. But today the public sector has lost
its cachet in the West. With a few small exceptions, the federal
government "does average" the whole time.

Continuing his drive against money politics, Bill Lincoln would
massively reduce the number of political appointees. Ambassadorships would no longer be for sale; they would go to diplomats.

He would start paying the heads of government departments the same sort of salaries they could get in the private sector—while imposing limits on what they can do after public service (to close the revolving door to K Street). He would copy Singapore's idea of scholarships for public service: pay the full fees of poor students at elite universities in return for a commitment that they will work for, say, five years in the public sector. Why shouldn't a bright kid from the Bronx go to Harvard for free but then spend a few years working as a public servant? This would allow the Ivy League to broaden its catchment and thereby reconnect it with the life of the nation (currently more Harvard students come from the top tenth of the population in terms of wealth than from the bottom half). It could also provide, say, the Department of Housing and Urban Development with access to talented young people who treat the citizens of the Bronx as brothers, cousins, and friends rather than just as statistics. Personnel is perspective as well as policy.

11. NATIONAL SERVICE FOR ALL

This points to one of the most troubling problems with modern America: the fraying of the bond between the elite and the public realm—the state. Rich Americans used to regard public service as a badge of honor. The first President Lincoln drafted soldiers from every class, as did his successors in the First and Second World Wars. As Teddy Roosevelt said, "To whom much has been given much shall be expected." There is mercifully no need for compulsory military service today. But America would gain enormously if every young man and woman was expected to work eighteen months for the government before the age of twenty-five, serving in some category or another.

Again, this is an idea that other countries are experimenting with, despite having much lower levels of inequality than America. Sweden reintroduced national service in 2018; South Korea is considering introducing "social service" as an alternative to military service. Emmanuel Macron has also raised the subject. For all the traumas and political complications of serving in the West Bank, national service in Israel has created a bond between the social classes and between the people and the state; it has also helped supercharge the tech economy.

In his inaugural address in 1960, John F. Kennedy issued a clarion call to the young: "Ask not what your country can do for you but what you can do for your country." Since then America has increasingly divided between "takers" who only ask what their country can do for them and "shirkers" who want to have as little to do with the government as possible. A civilian corps would once again force Americans to mix across class lines and help draw a fragmented country together. If Harvard's gilded youth were forced to dig roads and guard prisons alongside school dropouts from Compton and the Bronx, they might take more interest in both the roads and the dropouts. At a time when it is faltering in its competition with China, President Lincoln would revive the call.

12. MAKE GOVERNMENT DOWDY

Prime Minister Gladstone was so bent on saving money that he told his government to use cheaper writing paper. In Abraham Lincoln's time, Washington was a small southern town that was regarded as a hardship posting for diplomats (you got paid more to serve in the swampy summer). Now it has become one of the richest zip codes in the country, with Ritz Carltons, Tiffanys, Mc-Mansions, and Morton's steak houses spilling into the Virginia and Maryland suburbs. Politicians who used to rush back to their hometowns when they finished their political careers now routinely stay on in the capital as lobbyists and "advisers."

The presidency has become ever more imperial. Presidents sweep through Washington, DC, in giant motorcades. They fly around the world in a personalized jumbo jet. They take their own car, the Beast, with them wherever they go, which is no small feat given that it weighs ten tons. They hold elaborate multicourse dinners in the White House with guest lists that glitter with Hollywood stars (Deng Xiaoping once found himself sitting next to Shirley MacLaine who gave him a lecture on the virtues of the Cultural Revolution). They spend a lot of time playing golf with billionaires. This sends the wrong message to the people: that the government has plenty of money to spare.

We have to admit that, despite his reputation for parsimony, the first President Lincoln overspent on redecorating the White House. Our President Lincoln would copy Gladstone—living on bread and water, spending his evenings reading Homer, and walking everywhere. His modern model would be Pope Francis. The pontiff lives like a pauper and drives himself round in a simple old Renault. It is much easier for a pope or a president to cut unnecessary spending if you don't spend your life in a cocoon of privilege. Presidents shouldn't live like emperors.

13. REBUILD THE WEST AND EXPAND IT

One thing that Bill Lincoln would study is the "Huawei barometer," a map on Bloomberg's website that shows which countries accept the Chinese tech giant's equipment.[19] It is an accurate measurement of the decline in American soft power. So far, despite energetic arm-twisting, not to mention all the money it still spends on its allies' defense, the United States has persuaded only Australia, Japan, and Taiwan to ban Huawei. Britain may join them, but most of Europe has ignored America's pleas. So has much of democratic Asia. That is partly because Huawei's equipment is cheap and useful, but America's allies are also fed up with being ignored and insulted. The same goes for trade: if the United States, Europe, and Asia's democracies were to negotiate with China as one, they would be an unstoppable force. The West's collective failure to prevent China from breaking its promises in Hong Kong was another reminder of the weakness that disunity brings.

If the United States keeps on losing allies, its power will ebb, no matter how much reform it does at home. Bill Lincoln would start rebuilding Western institutions, especially NATO. And he would draw on the oratorical power of his nineteenth-century antecedents and sing the song of freedom. For this is not just about geopolitics. It is about liberty—and moral suasion. This used to be a big part of America's message. Sometimes it ended up looking naive or hypocritical. But fighting for freedom and democracy rather than just narrow national self-interest served America well during the Cold War. It will serve it once again as it tries to prevent China from making the world "safe for autocrats."

Our President Lincoln believes that the West is about more than just geography. Rather than looking through the prism of "America First" and nationalism, he would try to unite the democracies, so they speak as one. Rebuilding the ties with Europe is fundamental to that: why isn't there a free trade pact? But Bill Lincoln would not stop there. He would bring the democracies of Asia into the organizations of the West, so that countries like South Korea, Indonesia, and India (not to mention Japan and Australia) are defined by their freedoms, not their location. Trade is one obvious way. Reviving the Trans-Pacific Partnership that embraces twelve Pacific countries, including America, Canada, Chile, Singapore, and Japan, would make sense. The West needs to be expanded, not militarily but as a state of mind, so that when China tramples Hong Kong or other reversals of freedom happen, the democracies speak as one. Collectively they carry far more clout than the autocracies.

Reengaging with the global institutions that the United States helped found has to be part of any fightback. Many multilateral bodies are suffering from the same problems of old age as the federal government: bloat, self-obsession, and hypocrisy (the cause of human rights is hardly advanced by letting Saudi Arabia be head of the Human Rights Council). But that is a reason to re-engage with them, not shun them. With America absent, China is increasingly taking the lead. China's nationals now head four of the UN's institutions compared with just one American, and when Trump withdrew funds from WHO, the Chinese promised $2 billion. By all means point out the UN's failures. But also commit to its higher mission: show that America's heart does not stop at its borders. Send good people to the UN. Let those on national service work for global agencies too. For all China's clever "mask diplomacy" in the wake of the virus, there is still a deep distrust of the Middle Kingdom across the world, not least in Asia, and an affection for the United States. A little bit of sugar and a lot of pious words would go a long way.

THE LEARNING PRESIDENT

So those are the great causes that Bill Lincoln would undertake on behalf of the United States. Most of this agenda is merely copying what works elsewhere. There are plenty of examples of good government to choose from. If you don't like Canada's approach to health care, try a version of France's, which also uses public and private hospitals. America, in the tradition of declining empires, is bad at learning from other countries. Just as the Chinese emperor assumed Lord Macartney had nothing to show him, even when Britain was racing ahead, Washington seldom copies what works abroad. It has been talking about using Germany's system of technical training for generations; under Barack Obama, it flirted with copying Sweden's entitlement reforms. Like the proverbial boiling frog, the federal government may not discover how hot the water is until it is poured into a Chinese wok.

We sent Bill Lincoln to reform America, but you can apply the same thought experiment to any Western capital. If for instance he had been reborn as a Briton and installed in Downing Street, he would do some of the same things. National service would be helpful in binding Brexit Britain together; sunset clauses could help cut red tape; more power could be given to mayors; and the tax code could be simplified. Other things are different. Britain's pension system is on a sounder footing than America's. Health is the most difficult area, not least because British doctors demonize any attempt at reform as privatization. Covid certainly often showed the NHS at its best but it also highlighted problems, including control freakery and poor technology: just try Googling "National Health Service Covid contact tracing App." Many continental European countries have better health systems, without sacrificing anything in terms of coverage.

In continental Europe, Bill Lincoln would probably focus more on pensions and the social spending that Angela Merkel complains about. But he would be struck above all by the poor architecture of the European Union. If you want free movement of people, you need an external border force. As for the eurozone, the lesson from economics is simple: you cannot have a successful currency union without a banking union and a "transfer union." Northern Europeans think the latter means subsidizing the south, but less productive economies, like Italy and Greece, will always be at a disadvantage if they cannot devalue their currencies. Germany has been the great winner from the eurozone, just as New York and California win from being part of America's single market. Like its American peers, Germany should accept that transfers to the south are part of the deal, if it wants a common currency. Part of the deal could be southern Europe accepting reforms.

All across the West, democracies need to ask hard questions. Should governments offer state employees jobs for life? Should they restrict access to the teaching profession to people with teaching certificates? If video consultation for doctors worked so well during the crisis, why not embrace it in normal times? Should governments pay for students to go to university, when the gains from higher education go overwhelmingly to the privileged? Why hasn't the West invested in smart cities in the same way that Asia has?

Leviathan everywhere could be simplified and improved just by pragmatic modernization. You can get a long way just by repair work, room by room. But Covid gives us a chance to rethink the overall design, something which has not happened for decades. Here the argument changes from pragmatism to ideology, and technological improvement to political theory. What is the state for?

CONCLUSION:
MAKING GOVERNMENT GREAT AGAIN

I n 1969, with the ambitions of the federal government running out of control, Senator Daniel Patrick Moynihan issued a warning: "The stability of a democracy depends very much on the people making a careful distinction between what government can do and what it cannot do. To demand what can be done is altogether in order: some may wish such things accomplished, some may not, and the majority may decide. But to seek that which cannot be provided, especially to do so with the passionate but misinformed conviction that it can be, is to create the conditions of frustration and ruin."[1]

The West has generally failed to make the distinction about what government can do and what it can't. What does modern political theory tell us? The journalist John Authers has cleverly divided competing responses to the crisis into three broad philosophical camps: Rawlsians, libertarians, and communitarians.[2] The first group apply John Rawls's "veil of ignorance" test in *A Theory of Justice* (1971): what sort of society would you want if you didn't know whether you were going to end up rich or poor? In the battle with Covid, such thinking justifies free health care and job furlough schemes. Libertarians, who follow Robert Nozick's *Anarchy, State and Utopia* (1974), come at the problem from the other end. Normally reluctant to give the state any role beyond enforcing contracts and protecting against force, theft, and fraud, they have been suspicious of the powers that states have grabbed because of Covid. Communitarians, who lament the atomizing power of the market, have found their voice in the ground-up movements that evolved spontaneously during the pandemic—such as people clapping for health-care workers, or even taking out their sewing machines and making gowns for medical staff.

The difficulty comes when you try to jump from here to redesigning the state. The communitarians seem a bit naive: clapping for health-care workers is admirable, but it can't make up for Leviathan failing to impose quarantines. Far from bringing us together, plagues have often led to purges and pogroms. The libertarians seem unrealistic in a different way. The pandemic has reminded us that there are certain things that the state must do: providing medical research for vaccines, stepping in to prevent economic collapse, and rescuing society from selfish idiots such as the twenty-year-old student, partying on his spring break in Florida, who declared that "if I get corona, I get corona," indifferent to his impact on other people. The Rawlsians come closest to a redesign—but their enthusiasm for redistribution risks overloading an already overloaded state.

Modern political theory is bedeviled by ideological purity. The left has become fixated by equality of all sorts: it is hard to see how Isaiah Berlin, who believed that liberty was essentially about freedom from state compulsion, would have fitted in today's left, despite his fear of being branded a conservative. On the other side, too much thought on the right has gone into simply making the state smaller. Conservatives, supposedly the pragmatists, have often got more impractical: Michael Oakeshott had a lot to say about the case for resisting change (and drinking champagne) rather than dreaming up grandiose schemes. But when doing nothing means ever more rapid decline, or failing to prepare for pandemics and climate change, that does not help much.

The other bigger problem is that the theory of the state—particularly as a practical entity that runs things and makes trade-offs—has become something of a backwater. There has been a much greater focus on what individuals should do and on concepts like fairness, tradition, community, and personal ethics. Political theory often takes for granted that somewhere in the background there will be a broadly democratic welfare state. In 2016, the philanthropist Nicolas Berggruen, who has always had a keen interest in improving government, set up a million-dollar prize for Philosophy and Culture. The winners so far—Charles Taylor, Onora O'Neill, Martha Nussbaum, and Ruth Bader Ginsburg—are all significant thinkers, but none of them has focused on redesigning the state.

Of course this may change. University campuses are alive with political debate again—though not with arguments about what our contract with Leviathan should be. Until modern thought catches up, any answer will revolve around the three things that have dominated our story of the state—security, liberty, and leadership.

LEVIATHAN AND LOCKDOWNS

Begin with security. The Covid crisis shows Hobbes was right about the basic function of Leviathan. In 2020, government in most countries essentially became two departments—health and treasury. Liberal taboos against keeping soldiers off the streets disappeared. Ancient freedoms to roam at will—or, in Boris Johnson's formula, "to go to the pub"—were abandoned. But what do we mean by security once Covid recedes?

Our contract with the state has to mean more than just preventing egotistical citizens from killing each other, or protecting us from the plague. In the late nineteenth century, the case for building a welfare state was driven by fear of the spread of infectious diseases. Reformers like the Webbs argued that you needed to rebuild slum areas to prevent the disease from getting a foothold. Then you needed to provide better nutrition to improve people's ability to fight off infectious diseases. And then better schools and a decent food supply to make sure you had a fighting-fit, well-educated population.

The question now is where the national minimum should be set in a dynamic knowledge economy. We have already explained why America needs a national health service for practical reasons. There are philosophical reasons too. The contract between the richest country on the planet and its people surely entails providing free (or nearly free) medical care in the same way as it involves providing free (or nearly free) education. Franklin Roosevelt argued that government should provide security for "the forgotten man at the bottom of the economic pyramid." That man (or woman) may need more education later in life: and many people would classify free internet access as part of national infrastructure, rather like roads and bridges once were.

The difficulty comes when you jump from providing security (and equality of opportunity) to equality of outcomes. There is plainly no reason why the state should actively subsidize inequality: the tax breaks that the United States doles out to property developers and private equity break that rule, as do the protections that European countries give to (usually older, middle-class) workers in jobs, at the expense of (usually poorer, younger) people, stuck in unemployment or casualized labor. Leviathan should not boost the rich over the poor. But most attempts to create equality through forced redistribution have not only failed to create equality, they have suppressed liberty, dynamism, and ultimately prosperity.

Race is harder. Our basic sympathy is with the liberal ideal that everybody should be judged as an individual rather than as a member of a group (especially one you are born into). As far as possible, people should be allowed to be the authors of their own fates and the architects of their own identities. But liberal individualism has a problem: what about collective wrongs that have been imposed upon groups of people for no other reason than their skin color? Lyndon Johnson expressed this in a speech to Howard University: "You do not take a person who, for years, has been hobbled by chains and liberate him, bring him up to the starting line of a race and then say, 'you are free to compete with all the others,' and still justly believe that you have been completely fair." A collective injustice—such as the legacy of slavery and the continuation of racism in America—warrants collective remedies, as long as the ultimate goal clearly remains to treat everyone as individuals, rather than as members of a group. The target should be a race-blind society that can be race-blind precisely because it has eliminated the legacy of racism.

THE STATUTE OF LIBERTY

Security has to be balanced by liberty—and sometimes subordinated to it. The state needs to be constrained—not just by democracy, but also by checks and balances. The gaping hole in Hobbes's argument was that he didn't allow any room for people watching over Leviathan to make sure that it wasn't overreaching. John Stuart Mill believed that government needed to be monitored and restrained by a parliament and a free press.

Freedom is not only a good thing in itself for all individuals. It also promotes all sorts of wealth-producing and happiness-promoting virtues such as creativity and innovation that can improve the greater security of society. At the moment, our greater freedom to think and act accounts for the creative edge that the West holds over China. Mill thought that liberal democracy converts subjects into citizens. As he put it in *Considerations on Representative Government* (1861), the citizens of democracies become "conscious members of a great community" by engaging in political argument and discussion.[3] The citizens of autocracies, by contrast, are infantilized because they allow others to make their decisions for them. For all China's triumphs and America's failures, there are still more successful Chinese people who want to live in America than there are successful Americans who want to live in China.

Even before Covid, it was worth judging the West through this lens too. In the age of money politics, low turnout at elections, and rules passed for special interests that nobody else understands, how many of us are really "conscious members of a great community"? Simplifying and cleansing Leviathan would promote participation and thus freedom. Now Covid makes Mill's strictures about an overmighty state destroying liberty even more urgent.

In order to prevent people from spreading a potentially deadly disease, the state certainly has the right to restrict our freedoms and monitor our lives. We should be prepared to give up more information about ourselves than we have in the past, but only if it makes us significantly safer. And all this should be both conditional and temporary. Hobbes's citizens gave their rights to Leviathan forever; we should not. Security has always been the excuse for tyranny. Liberty and privacy matter, even if they sometimes bring a little more danger with them. When it comes to safety, even the best-intentioned governments never know when to stop. Leviathan gathering data to prevent an epidemic does not justify it gathering data to make everybody slimmer or ensure that people's feelings aren't hurt by political incorrectness.

A particular horror to bear in mind is the Panopticon. Mill's godfather, Jeremy Bentham, designed the perfect prison: circular in structure, which allowed a prison warder to keep watch on his charges from a central lodge, without them seeing him. Bentham's logic was that the prisoners would feel under constant inspection, even when they weren't. He wanted them to be kept in solitary confinement almost all the time and forced to wear masks for communal meals (so they couldn't communicate with each other). Although its construction was never completed, the Panopticon has got a new lease of life in the way modern society measures, classifies, and regulates its citizenry. Michel Foucault cited the Panopticon in *Discipline and Punish* (1975), when he worried about the "disciplinary society" that controls people more through invisible observation than through direct physical punishment.

China's use of facial recognition technology is designed to make a Panopticon out of the entire country. Is the West slipping into the same trap? Google and Facebook make their money by monitoring their customers, providing them with free and useful services but also raising the specter of what Shoshana Zuboff has called the "Age of Surveillance Capitalism." The Western state is collecting ever more data on us, somewhat chaotically. Covid has given the state an excuse to create a surveillance society. Israel has even authorized Shin Bet, its domestic security force, to break into people's mobile phones without their permission. The Panopticon may help keep us alive, but it is also bringing us closer to a future in which we are watched by our smartphones, filmed by cameras on every street corner, and obliged to scan bar codes when we get on the train.

There should be a balance not just between security and liberty but between utility and liberty. Just because something is useful, it does not mean we should give away freedoms to get it. At the moment the main danger is of a Panopticon by stealth, where the defaults are all set at "hand over my data." Liberals should step in to stop this. In Mill's day, liberals objected to opening the mail of bomb-throwing anarchists, because they thought the right of privacy was important: what would they have said about people handing over their entire life history just to get cheap email? Especially if they did not realize that was the deal being offered in the small print? Before Covid there was already a compelling case for much clearer privacy guarantees; the need is even greater today.

This reminds us of the oldest problem in democracy: who watches the watchers? Part of the answer lies in checks and balances. When spies want to eavesdrop on people, they need oversight. In some cases, consumers need to be told they should say no to something that is useful. One of the frightening things about the populists' attacks on the judiciary and the press is their scorn for the idea of balance. The will of the majority should not run roughshod over basic rights. But you cannot just rely on constitutional arrangements or legal precedence to protect you. You also need a cadre of people who care about ancient liberties, who are prepared to stand up for minority rights, and who take government and governance seriously. They seem to be missing at the moment.

THE NEED FOR GUARDIANS

Plato was fond of comparing the art of government to the art of navigation. The ship of state can easily be blown off course and either sent in the wrong direction or wrecked on the rocks. Plato asks a simple question: should we give the job of steering the ship safely to the members of the crew who don't know what they're doing, where they're going, or the rudiments of navigation? Or should we give it to the captain who has spent his entire life studying "the seasons of the year, the sky, the stars, the winds and other professional subjects"?

Plato's *Republic* hammers home two arguments that are highly relevant today. The first is that good government is vital. Fail the good government test—either by giving too much power to dictators or to the mob or to the rich—and you are doomed to misery. The second is that good government depends on the quality of the leadership elite. According to Plato, the state's most important job is to spot potential leaders when they are still young and provide them with an education that both stretches their abilities and, even more importantly, inculcates them with a sense of public service. These would be the guardians of the state.

It should be said that not all of Plato's arguments have stood the test of time. He was so keen on getting the best guardians that he advocated eugenic breeding programs and banned potential leaders from holding property of their own or even getting married (allowing them as a compensation to participate in regular orgies with women selected for their brains and beauty). He was also far too critical of democracy. But his starting point—that statecraft matters immensely—is something that many parts of the West have lost.

The big difference between the countries that have done well with Covid and those that have not is simply that they have taken government seriously. They have studied the art of government and modernized their states. They have also inculcated a sense of public duty and guardianship. It is difficult to imagine Plato selecting Donald Trump as the captain of his ship. By contrast, Angela Merkel has always known that government matters. Brought up in East Germany, she knows what tyranny means. Trained as a scientist, she knows how to evaluate evidence. She has never been a particularly visionary politician. She has also often been guilty of changing course too slowly. But this is a serious woman who is doing serious work.

FOR WHOM THE BELL TOLLS

The Athens of Plato and Pericles was not perfect. But in the ancient world it was nevertheless the closest thing to what the West represents today: there was a degree of democracy, freedom of thought, and astonishing bursts of creativity. It was also far preferable to Sparta—a thuggish military dictatorship that left babies to die if they showed any sign of weakness and lived for conquest. Athens lost because it was weakened by the plague and divided by internal disagreements.

For all its imperfections, liberal democracy is the best attempt humankind has come up with to solve those tensions between security and liberty and between freedom and order. The alarm clock has rung loudly. Will the West wake up in time? If this book has inspired, provoked, or infuriated at least a few people to start taking government seriously again, then it will have been worth writing.

ACKNOWLEDGMENTS

Producing a book quickly, albeit a short one, means that even more than usual we have relied on the kindness of others. We would like to thank Andrew Wylie and James Pullen at the Wylie Agency for selling our idea; and Judith Curr, Juan Milà, Aurea Carpenter, and Catherine Gibbs not just for embracing it but being such patient editors. In particular, this project would have remained a mad idea without Rebecca Nicolson. Perhaps it still is—but that is not her fault.

We would like to thank Mike Bloomberg and Zanny Minton-Beddoes, the editor of *The Economist*, for giving us permission to do it. Reto Gregori, David Shipley, Nick Boles, Graham Mather, and Fredrik Erixon all read drafts and made incredibly helpful suggestions. Rosie Blau and Fani Papageorgiou shared their knowledge of China and Greece respectively. Gideon Rachman remained an inspiration. Mark Doyle once again helped us research it. Thank you to all of them. All errors, though, remain our own.

As usual, the main burden has fallen on those we live with. Normally books are an excuse to get rid of us—at least for a while. Lockdown removed even that compensation. We would like to thank Kristin, Tom, Guy, and Eddie, and Amelia, Ella, and Dora for their tolerance.

London, July 2020

NOTES

INTRODUCTION: WEEKS WHEN DECADES HAPPEN

1. We have used numbers throughout from the Bloomberg virus tracker, which is largely based on Johns Hopkins research.
2. Shalini Ramachandran, Laura Kusisto, and Katie Honan, "How New York's Coronavirus Response Made the Pandemic Worse," *Wall Street Journal*, June 11, 2020.
3. Chris Morris and Oliver Barnes, "Coronavirus: Which Regions Have Been Worst Hit?" BBC News, June 3, 2020.
4. George Packer, "We Are Living in a Failed State," *The Atlantic*, June 2020.
5. Philip Wen and Drew Hinshaw, "China Asserts Claim to Global Leadership, Mask by Mask," *Wall Street Journal*, April 1, 2020.
6. *Wall Street Journal*/NBC poll, June 7.
7. Boyd Hilton, *A Mad, Bad, and Dangerous People? England 1783–1846* (Oxford: Oxford Univ. Press, 2006), 558.
8. Alice Tidey, "UK Records More New Covid-19 Deaths Than Entire EU Combined," *Euronews*, June 4, 2020.
9. Caroline Wazer, "The Plagues That Might Have Brought Down the Roman Empire," *The Atlantic*, March 16, 2016.

CHAPTER ONE: THE RISE OF THE WEST

1. Michael Massing, *Fatal Discord: Erasmus, Luther, and the Fight for the Western Mind* (New York: Harper, 2018), 258.
2. Hilary Mantel, *The Mirror & the Light* (London: Fourth Estate, 2020), 366.
3. Some historians think He's treasure ships were 120 meters long; others say no more than 80 meters. Da Gama's flagship was 27 meters long.
4. David Landes, *The Wealth and Poverty of Nations: Why Some Are So Rich and Some So Poor* (New York: W. W. Norton & Company, 1998), 35.
5. Landes, *The Wealth and Poverty of Nations*, 36, 38.
6. Jonathan Spence, *The Search for Modern China* (New York: W. W. Norton, 1999), 122–23.

7. Anthony Trollope, *The Way We Live Now* (London: Penguin ed., 1994), 545.

8. G. R. Elton, *Reformation Europe 1517–1559* (London: Fontana, 1979), 298–99.

9. John Micklethwait and Adrian Wooldridge, *The Fourth Revolution: The Global Race to Reinvent the State* (New York: Penguin Press, 2014), 34.

10. Charles Tilly, "Reflections on the History of European State Making," in *The Formation of National States in Western Europe*, ed. Charles Tilly (Princeton: Princeton Univ. Press, 1975), 42.

11. Geoffrey Parker, *The Military Revolution: Military Innovation and the Rise of the West 1500–1800* (Cambridge: Cambridge Univ. Press, 1998).

12. Charles Spencer, *To Catch a King: Charles II's Great Escape* (London: William Collins, 2017), 41.

13. Alan Ryan, *On Politics: A History of Political Thought from Herotodus to the Present* (London: Allen Lane, 2012), 445–46.

14. Francis Fukuyama, *The Origins of Political Order: From Prehuman Times to the French Revolution* (London: Profile Books, 2011), 124.

15. Jonathan Israel, *Radical Enlightenment: Philosophy and the Making of Modernity* (Oxford: Oxford Univ. Press, 2001), 2–3.

16. Thomas Paine, *Common Sense*, 1776, Project Gutenberg e-book.

17. Andrew Gimson, *Gimson's Prime Ministers: Brief Lives from Walpole to May* (London: Square Peg, 2018), 135.

18. J. S. Mill, *Autobiography*, 156, Project Gutenberg e-book (first published 1874, this edition 2003).

19. Boyd Hilton, *A Mad, Bad and Dangerous People? England 1783–1846* (Oxford: Oxford Univ. Press, 2006), 558.

20. Speech to the House of Commons, March 19, 1850; https://www.gladstoneslibrary.org/news/volume/a-statement-from-gladstones-library-black-lives-matter.

21. Gimson, *Gimson's Prime Ministers*, 133–34.

22. William Ewart Gladstone (by H.C.G. Matthew), *Dictionary of National Biography*, 22:389.

23. Martin Daunton, *State and Market in Victorian Britain: War, Welfare and Capitalism* (Woodbridge: Boydell Press, 2008), 73–74.

24. A. V. Dicey, *Lectures on the Relations Between Law and Opinion in England During the Nineteenth Century* (London: Macmillan, 1920), 430–31.

25. Bertrand Russell, *The Autobiography of Bertrand Russell, 1872–1914* (London: Allen & Unwin, 1967), 1:107.

26. Quoted in W. H. G. Armytage, *Four Hundred Years of English Education* (Cambridge: Cambridge Univ. Press, 1970), 174.

27. Vito Tanzi, *Government versus Markets: The Changing Economic Role of the State* (Cambridge: Cambridge Univ. Press, 2011), 126.

28. Quoted in Thomas Frank, *The Wrecking Crew: How Conservatives Rule* (New York: Metropolitan Books, 2008), 15.

29. *The Economist*, November 19, 1955.

30. G. Calvin Mackenzie and Robert Weisbrot, *The Liberal Hour: Washington and the Politics of Change in the 1960s* (New York: Penguin Press, 2008), 357.

31. Mackenzie and Weisbrot, *The Liberal Hour*, 313.

32. Alan Greenspan and Adrian Wooldridge, *Capitalism in America: A History* (New York: Penguin Press, 2018), 303.

CHAPTER TWO: THE DECLINE OF THE WEST

1. Edward Gibbon, *Autobiography* (Letchworth, UK: Temple Press, 1924 ed.), 126.

2. Christian Caryl, *Strange Rebels: 1979 and the Birth of the 21st Century* (New York: Basic Books, 2013), 183.

3. Charles Moore, *Margaret Thatcher: The Authorized Biography* (London: Allen Lane, 2013), 1:315.

4. Steven F. Hayward, *The Age of Reagan: The Fall of the Old Liberal Order, 1964–1980* (New York: Forum, 2001), 321.

5. State of the Union address to Congress, January 19, 1978.

6. Caryl, *Strange Rebels*, 160.

7. Moore, *Margaret Thatcher*, 1:352.

8. Charles Moore, "The Invincible Mrs. Thatcher," *Vanity Fair*, November 2011.

9. Daniel Yergin and Joseph Stanislaw, *The Commanding Heights: The Battle Between Government and the Marketplace That Is Remaking the Modern World* (New York: Simon & Schuster, 1998), 123.

10. Peter Schuck, *Why Government Fails So Often. And How It Can Do Better* (Princeton: Princeton Univ. Press, 2014), 318.

11. "Taming Leviathan: A Special Report on the Future of the State," *The Economist*, March 19, 2011, 5.

12. House Small Business Subcommittee on Contracting and Workforce's hearing on "The Decline in Business Formation: Implications for Entrepreneurship and the Economy," September 11, 2014, 66.

13. Edward McBride, "Cheer Up: A Special Report on American Competitiveness," *The Economist*, March 16, 2013.

14. Brad Hershbein, David Boddy, and Melissa Kearney, "Nearly 30 Percent of Workers in the US Need a License to Perform Their Job," Brookings Institution, January 27, 2015.

15. Thomas Frank, *The Wrecking Crew: How Conservatives Rule* (New York: Metropolitan Books, 2008), 37.

16. Jonathan Hopkin, *Anti-System Politics: The Crisis of Market Liberalism in Rich Democracies* (Oxford: Oxford Univ. Press, 2020), 115.

17. George Brown, Britain's foreign secretary under Harold Wilson.

18. Graham Allison and Robert D. Blackwell, with Ali Wyrne, *Lee Kuan Yew: The Grand Master's Insights on China, the United States and the World* (Cambridge, MA: MIT Press, 2013), 32.

19. Grace Ho, "No Mid-Year Bonus for Civil Servants; One-Time Pay Cut for Superscale Public Officers," *Straits Times*, June 18, 2020.

20. Evan Osnos, *The Age of Ambition: Chasing Fortune, Truth and Faith in the New China* (London: The Bodley Head, 2014), 66.

21. Hopkin, *Anti-System Politics*, 63.

22. George Packer, "How to Destroy a Government," *The Atlantic*, April 2020.

CHAPTER THREE: THE OVERLOADED STATE

1. Michael Garland and Gaurav Pal, "Government Needs to Get Serious About IT Modernization," *The Business of Federal Technology*, June 18, 2019.

2. Paper from the Partnership for Public Service: Tech Talent for 21st Century Government, April 2020.

3. Statement of Max Stier to the House Committee on Oversight and Government Reform's hearing on "Workforce for the 21st Century," May 16, 2018.

4. This is helped, it should be disclosed, by Bloomberg Philanthropies.

5. Alan Greenspan and Adrian Wooldridge, *Capitalism in America: A History* (London: Penguin, 2018), 404–5.

6. Morris Kleiner, "Reforming Occupational Licensing Policies," *The Hamilton Project*, March 2015.

7. Mancur Olson, *The Logic of Collective Action: Public Goods and the Theory of Groups* (Cambridge, MA: Harvard Univ. Press, 1965), 36.

8. Peter Schuck, *Why Government Fails So Often: And How It Can Do Better* (Princeton: Princeton Univ. Press, 2014), 322.

9. Troy Senik, "The Worst Union in America," in *The Beholden State: California's Lost Promise and How to Recapture It*, ed. Brian Anderson (Boulder: Rowman and Littlefield, 2013), 203–5.

10. Schuck, *Why Government Fails So Often*, 175.

11. We should confess that one of us comes from a family of British farmers.

12. Selam Gebrekidan, Matt Apuzzo, and Benjamin Novak, "The Money Farmers: How Oligarchs and Populists Milk the EU," *New York Times*, November 3, 2019.

13. Colin Grubak, Inu Manak, and Daniel Ikenson, "The Jones Act: A Burden America Can No Longer Bear," *Cato Institute Policy Analysis*, June 28, 2018.

14. Schuck, *Why Government Fails So Often*, 177, 180.

15. We are indebted to Mario Calvo-Platero for this insight.

16. Group of Thirty, *Fixing the Pensions Crisis: Ensuring Lifetime Financial Security* (Washington, DC, 2019).

17. "Policy Basics: Where Do Our Federal Tax Dollars Go?" Center on Budget and Policy Priorities, April 9, 2020.

18. Greenspan and Wooldridge, *Capitalism in America*, 407.

19. Schuck, *Why Government Fails So Often*, 319.

20. "Law School Popular for Congress, with Harvard, Georgetown Topping List," *Bloomberg Law*, January 25, 2019.

21. Jeremy Paxman, *The Political Animal: An Anatomy* (London: Michael Joseph, 2002), 206–7.

CHAPTER FOUR: THE COVID TEST

1. John Burns-Murdoch and Chris Giles, "UK Suffers Second-Highest Death Rate from Coronavirus," *Financial Times*, May 28, 2020.

2. Tim Ross and Kitty Donaldson, "Boris Johnson Revamps Agenda to Meet Worst UK Recession in 300 Years," *Bloomberg*, June 2, 2020.

3. Marc Champion, "Coronavirus Is a Stress Test Many World Leaders Are Failing," *Bloomberg*, May 22, 2020.

4. Lara Zhou and Keegan Elmer, "Thousands Left Wuhan for Hong Kong, Bangkok, Singapore or Tokyo Before Lockdown," *South China Morning Post*, January 27, 2020.

5. Richard Horton, *The Covid-19 Catastrophe: What's Gone Wrong and How to Stop It Happening Again* (Cambridge: Polity Press, 2020), 53.

6. "New World Curriculum," *The Economist*, March 7, 2020, 19.

7. Jamie Grierson, "UK Government Under Fire After 'Big Influx' of Covid-19 Cases from Europe Revealed," *Guardian*, May 5, 2020.

8. Laura Donnelly, "Earlier Lockdown Could Have Prevented Three-Quarters of UK Coronavirus Deaths, Modelling Suggests," *Daily Telegraph*, May 20, 2020.

9. Rafaela Lindeberg, "Man Behind Sweden's Controversial Virus Strategy Admits Mistakes," *Bloomberg*, June 3, 2020.

10. Sen Pei, Sasikiran Kandula, and Jeffrey Shaman, "Differential Effects of Intervention Timing on COVID-19 Spread in the United States," posted on medRxiv preprint server, May 29, 2020.

11. Drew Armstrong et al., "Why New York Suffered When Other Cities Were Spared by Covid-19," *Bloomberg*, May 28, 2020.

12. Armstrong et al., "Why New York Suffered When Other Cities Were Spared by Covid-19."

13. Fareed Zakaria, "If New York Founders It Will Be Because of Bad Government, Not the Pandemic," *Washington Post*, June 11, 2020.

14. Raphael Rashid, "Being Called a Cult Is One Thing, Being Blamed for an Epidemic Is Quite Another," *New York Times*, March 9, 2020.

15. Laura Spinney, "The Coronavirus Slayer! How Kerala's Rock Star Health Minister Helped Save It from Covid-19," *Guardian*, May 14, 2020.

16. "New World Curriculum," *The Economist*.

17. Shalini Ramachandran, Laura Kusisto, and Katie Hoanan, "How New York's Coronavirus Response Made the Pandemic Worse," *Wall Street Journal*, June 11, 2020.

18. Full disclosure: Paul Deighton is chairman of *The Economist*.

19. Gordon Lubold and Paul Vieira, "US Drops Proposal to Put Troops at Canadian Border," *Wall Street Journal*, March 26, 2020.

20. Teresa Coratella, "Whatever It Takes: Italy and the Covid-19 Crisis," European Council on Foreign Relations, March 18, 2020.

21. Andy Hoffman, "Sex Workers Can Get Back to Business in Switzerland, but Sports Remain Prohibited," *Bloomberg*, May 20, 2020.

22. YouGov, "Americans Trust Local Governments over the Federal Government on COVID-19," April 27, 2020.

23. John Lichfield, "Coronavirus: France's strange defeat," *Politico*, May 8, 2020.

24. Anne Applebaum, "The Coronavirus Called America's Bluff," *The Atlantic*, March 15, 2020.

CHAPTER FIVE: THE MORBID SYMPTOMS

1. "The Right Medicine for the World Economy," *The Economist*, March 7, 2020.
2. Song Luzheng, "Many Western Governments Ill-Equipped to Handle Coronavirus," *Global Times*, March 15, 2020.
3. Iain Marlow, "China Trolls US over Protests After Trump Criticized Hong Kong," *Bloomberg*, June 1, 2020.
4. "Protection Racket," *The Economist*, April 25, 2020.
5. Richard Pérez-Peña, "Virus Hits Europe Harder Than China. Is That the Price of an Open Society?" *New York Times*, March 19, 2020.
6. Gary Bass, "The Athenian Plague: A Cautionary Tale of Democracy's Fragility," *New Yorker*, June 10, 2020.
7. Doug Klain, "Azerbaijan's Strongman Senses Opportunity in Coronavirus Pandemic," Atlantic Council, March 19, 2020.
8. Marc Champion, "Coronavirus Is a Stress Test Many World Leaders Are Failing," *Bloomberg*, May 22, 2020.
9. "Leviathan Rising," *The Economist*, March 21, 2020, 24.
10. Sebastian Mallaby, "The Age of Magic Money," *Foreign Affairs*, May 29, 2020.
11. June 18, 2020.
12. "The Fiscal Response to the Economic Fallout from the Coronavirus," Bruegel Datasets, May 27, 2020.
13. Isabel Reynolds and Emi Urabe, "Japan to Fund Firms to Shift Production Out of China," *Bloomberg*, April 8, 2020.
14. Catarina Saraiva, "Unrest Spotlights Depth of Black Americans' Economic Struggle," *Bloomberg*, June 2, 2020.
15. Public Health England, "Disparities in the Risk and Outcomes of COVID-19," June 2020.
16. Gerald Seib and John McCormick, "Coronavirus Means the Era of Big Government Is . . . Back," *Wall Street Journal*, April 26, 2020.
17. "Strategic Pile-Up," *The Economist*, April 11, 2020.
18. Mallaby, "The Age of Magic Money."

CHAPTER SIX: WHAT WOULD BILL LINCOLN DO?

1. Samuel Finer, *The History of Government from the Earliest Times*, vols. 1–3 (Oxford: Oxford Univ. Press, 1997–99). Finer was the Gladstone professor of government at Oxford University.

2. Chris Edwards, "Coronavirus and NIH/CDC Funding," Cato Institute, March 16, 2020.

3. Paul Collier, "The Problem of Modelling," *Times Literary Supplement*, April 24, 2020.

4. Environmental and Energy Study Institute, "Fossil Fuel Subsidies: A Closer Look at Tax Breaks and Societal Costs," July 29, 2019.

5. "Order Above the Law," *The Economist*, June 6, 2020.

6. Hugo Miller, "Europe Released 120,000 Inmates to Control Coronavirus," *Bloomberg*, June 18, 2020.

7. Pew Research Center, Fact-tank, April 30, 2019.

8. The figure is taken from the Tax Policy Center's Briefing Book.

9. "America's Public-Sector Pension Schemes Are Trillions of Dollars Short," *The Economist*, November 14, 2019.

10. "JCT Estimates Record $1.6 Trillion in Tax Breaks in 2017," Committee for a Responsible Federal Budget, March 28, 2017.

11. Tax Policy Center Briefing Book: How Could We Improve the Federal Tax System?

12. Dag Detter, *Unlocking Public Wealth: Governments Could Do a Better Job of Managing Their Assets* (Washington: The International Monetary Fund), 2018.

13. Garett Jones, *10% Less Democracy: Why You Should Trust Elites a Little More and the Masses a Little Less* (Stanford: Stanford Univ. Press, 2020), 4.

14. David Himmelstein, Terry Campbell, and Steffie Woodhandler, "Health Care Administrative Costs in the United States and Canada, 2017," *Annals of Internal Medicine*, January 21, 2020.

15. OECD, PISA results for 2018, released December 3, 2019; https://www.oecd.org/pisa/publications/pisa-2018-results.htm.

16. Department of Health and Human Services, "Strategic Objective 5.3: Optimize Information Technology Investments to Improve Process Efficiency and Enable Innovation to Advance Program Mission Goals," February 25, 2019, https://bit.ly/2WLdL8W.

17. "Shanghai Introduces QR Codes on Subway to Track Potential Contact with Coronavirus," *South China Morning Post*, February 28, 2020.

18. Aditi Kumar and Eric Rosenbach, "Could China's Digital Currency Unseat the Dollar?" *Foreign Affairs*, May 20, 2020.

19. Bloomberg: The Huawei barometer graphic.

CONCLUSION: MAKING GOVERNMENT GREAT AGAIN

1. Daniel P. Moynihan, *Coping: Essays on the Practice of Government* (New York: Random House, 1973), 255–56.

2. John Authers, "How Coronavirus Is Shaking Up the Moral Universe," *Bloomberg*, March 29, 2020, and "The Golden Rule Is Dying of Covid," *Bloomberg*, May 30, 2020.

3. John Stuart Mill, *Three Essays* (Oxford: Oxford Univ. Press, 1975), 275–76.

Here ends John Micklethwait and Adrian Wooldridge's
The Wake-Up Call.

The first edition of this book was printed and
bound at LSC Communications in
Harrisonburg, Virginia, August 2020.

A NOTE ON THE TYPE

The text of this book was set in FF Tisa, a serif typeface
created by Slovenian designer Mitja Miklavcic in 2006.
Miklavcic, then a graduate student at the University of
Reading, drew inspiration from nineteenth-century slab
serif typefaces. FF Tisa maintains the authority of the slab
serif genre while improving legibility and elegance with
its pronounced serifs and slightly exaggerated ink traps.

HARPERVIA

An imprint dedicated to publishing international voices,
offering readers a chance to encounter other lives and other
points of view via the language of the imagination.